2

THE HERMES EXPERIMENT

A Personal Story

THE HERMES EXPERIMENT

A Personal Story

Richard Milner

Massachusetts Institute of Technology, USA

Erhard Steffens

University of Erlangen-Nürnberg, Germany

World Scientific

NEW JERSEY · LONDON · SINGAPORE · BEIJING · SHANGHAI · HONG KONG · TAIPEI · CHENNAI · TOKYO

Published by

World Scientific Publishing Co. Pte. Ltd.

5 Toh Tuck Link, Singapore 596224

USA office: 27 Warren Street, Suite 401-402, Hackensack, NJ 07601

UK office: 57 Shelton Street, Covent Garden, London WC2H 9HE

British Library Cataloguing-in-Publication Data
A catalogue record for this book is available from the British Library.

THE HERMES EXPERIMENT
A Personal Story

ISBN 978-981-121-533-9 (hardcover)
ISBN 978-981-121-534-6 (ebook for institutions)
ISBN 978-981-121-535-3 (ebook for individuals)

For any available supplementary material, please visit
https://www.worldscientific.com/worldscibooks/10.1142/11692#t=suppl

Desk Editor: Nur Syarfeena Binte Mohd Fauzi

Typeset by Stallion Press
Email: enquiries@stallionpress.com

To Eileen and Heike,
our loving and supportive partners in all we have accomplished

Preface

Understanding the fundamental structure of matter has been an important and unifying goal of human endeavor since the dawn of civilization. In the latter half of the twentieth century, physicists have developed a fundamental theory of the strong force, which binds the proton and nuclei, and which fuels the stars in our universe. This theory is based on unobservable particles: fractionally charged, point-like *quarks* exchanging uncharged, massless *gluons*, and it is consistent with the results of all laboratory experiments. However, describing basic properties of the proton, like its mass and spin, in terms of the fundamental theory has proved elusive. Surprising results from CERN in the late 1980s underscored that our understanding of the origin of the proton's spin was very incomplete. Thus, physicists in Europe and N. America conceived in 1988 a new, technically innovative experiment, called **HERMES** (**HERA ME**asurements of **S**pin-dependent Structure Functions of the Neutron and Proton), at the DESY electron–proton collider HERA to probe the origin of the spin of the proton and neutron. The HERMES experiment acquired data from 1995 to 2007, provided new insight into the quark and gluon structure of matter, and has shaped our modern understanding of the proton's fundamental structure. Two physicists who played a central role in originating and realizing the HERMES experiment provide a personal account of how the technically challenging experiment was conceived, successfully constructed and operated at the HERA collider in Hamburg, Germany, and utilized in ways unforeseen when initially proposed. Further, they summarize the considerable scientific impact of HERMES and describe how it has influenced our view of the fundamental structure of matter, a subject that continues to excite physicists worldwide with its scientific opportunities.

In this book, we describe our personal story concerning the origin, realization, running and scientific output of the HERMES experiment. We have made a significant effort to accurately paint the broader picture and have profited from critical reading and comments by our esteemed colleagues (see Acknowledgments). However, all remaining limitations in the book are the sole responsibility of the authors. The book is written for the interested reader with a non-technical background and lacks the rigor and completeness of a scientific review. Like all personal recollections, our perspective is necessarily incomplete.

We suggest that the reader use the book depending on their background. For the reader who comes to learn about HERMES for the first time, we have provided in the first two chapters an introduction and a context with which to understand the exciting science that has motivated the HERMES experiment. In addition, we have added Appendix A to explain units, scientific notation and common technical terms and Appendix B to explain the frequently used acronyms. For readers familiar with the material of the first two chapters, the reader can first read Appendix C on the history of spin and then proceed to Chapter 3, where the HERMES story begins. We hope that the curious reader will find value in our narrative and develop an appreciation for how a large scientific experiment can be conceived and realized. Finally, we hope that future generations of young scientists can find inspiration in the HERMES story as they work to address the profound questions they face in seeking to understand the physical world.

March, 2021

Richard Milner
Arlington,
Massachusetts, USA

Erhard Steffens
Weisendorf,
Bavaria, Germany

Acknowledgments

We gratefully acknowledge valuable conversations with Moscow Amarian, Elke Aschenauer, Harut Avakian, Jo van den Brand, Ed Kinney, Costas Papanicolas, Caroline Riedl, Klaus Rith and Gunar Schnell. We acknowledge the careful reading of the draft manuscript by Klaus Rith, Gunar Schnell and Albrecht Wagner, who provided valuable comments. We thank Ricardo Alarcon, Abhay Deshpande, and Rolf Ent for valuable comments on Chapter 6. We thank David Milner for his expertise in preparing the figures in the book and for his design of the cover.

RM thanks Willy Haeberli, Chris Keith, Alan Krisch, Matt Poelker, Charles Prescott, and Erhard Steffens for providing him with essential information and insight into the history of spin.

Contents

Chapter 1

Quest to Understand the Fundamental Structure of Matter

1.1 Structure of Matter

This book is the personal story of two physicists who played central roles in the realization of an innovative experiment, called HERMES, with the aim of gaining new understanding into the fundamental structure of matter. HERMES was conceived in the late 1980s, proposed and constructed by an international collaboration from Europe and North America, first installed in the East Hall of the HERA accelerator in Hamburg, Germany in 1995 and took data until 2007, when operations at HERA ceased. Its primary scientific motivation was to understand the origin of the spin of the proton and neutron, the fundamental building blocks of nuclei.

For the reader coming to the story for the first time, the HERMES experiment illustrates the creative ability of scientists to use novel techniques and large facilities in a serendipitous way to address fundamental questions. Because of the scale of human endeavor and large resources required, frontier research in subatomic physics is intrinsically programmatic. HERMES is an example where modest resources can leverage enormous investments in technical infrastructure to great effect.

In the first two chapter, we explain the meaning and importance of spin, paint the modern picture of the structure of matter for the reader and we set the stage for the motivation and discussion of HERMES.

1.1.1 *Atoms: The building blocks of matter*

The building block of the visible matter in the universe around us is the atom. An atom consists of a distribution of negatively charged electrons bound to a nucleus of equal positively charged protons and uncharged neutrons. An atom is about a million times smaller than a typical human hair. A typical nuclear size is about 100,000 times smaller than the atom. [Note that a scale reduction of 100,000 arises in going from your height to about the thickness of household aluminum foil.] Electrons are point-like objects with no structure, as far as we know, while protons and neutrons are extended objects made up of electrically charged constituents. The proton and electron have positive and negative charges, respectively, of equal magnitude and the neutron is electrically neutral. The masses of the proton and neutron are about 1,800 times the mass of an electron. The nucleus is built from the protons and neutrons via the strong force, described below. An atom in its ground state is always electrically neutral. Figure 1.1 shows schematically the structure of the atom.

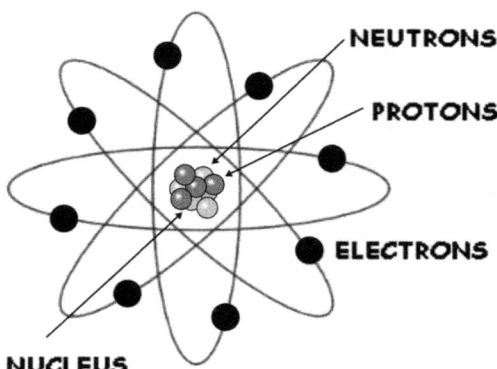

Fig. 1.1: Schematic layout of the structure of the atom showing the orbital electrons and the nucleus with its constituent protons and neutrons. In this figure, the size of the nucleus is increased by a factor of about 100,000 to make it visible. Further, the electrons are point particles. In reality, most of the atom consists of empty space occupied by the orbital electrons bound by the relatively tiny nucleus.

Atoms can share electrons which leads to chemical bonds. Understanding the structure and properties of bound atoms is the concern of chemists. Chemical reactions produce rearrangements of the electrons that involve energy changes. Atomic electrons like to order themselves in shells of a common energy. Closed electron shells are tightly bound, i.e. they are unreactive chemically, like those of noble gases. The energy scales in chemical reactions are of the order of the energy levels in atoms while those involving nuclear reactions are about 1 million times larger. For example, the Sun, like all stars, is fueled by nuclear reactions but the resulting sunlight reaching Earth is responsible for essential chemical reactions like photosynthesis. See Appendix A for more details.

Free neutrons, being slightly heavier, can decay via the weak force into protons, but the inverse is forbidden.

1.1.2 *Mysterious quantum world*

Quantum mechanics was developed in the early decades of the twentieth century to explain the atom and its constituents. Objects can be thought of as being simultaneously both particles and waves at the quantum level. Physical systems have wave functions which are produced as solutions of basic equations. The wave function is not observable directly. However, its square can be interpreted as a probability distribution. In quantum mechanics, outcomes have probabilities and lack the definiteness we associate with physics in the everyday world we live in (see Fig. 1.2, showing E. Schrödinger, the originator of the wave function).

An important result is the *Heisenberg Uncertainty Principle*. This states that one cannot know with arbitrary precision simultaneously the position and momentum of a quantum system. Practically, it means that the constituents of the proton and neutron must have significant momentum, as they are confined to a small region in space. Similarly, the protons and neutrons which form a nucleus have significant momentum. Everything is moving in the quantum world of the atom. Another consequence of the Heisenberg Uncertainty Principle is that the energy of a subatomic particle cannot be

Fig. 1.2:   The Austrian physicist Erwin Schrödinger, who originated a formulation of quantum mechanics in which the operators do not vary with time.

determined with arbitrary precision at a given point in time (see Fig. 1.3, showing W. Heisenberg, famous for his uncertainty relation).

Another fundamental aspect of quantum mechanics, which is central to our story here, is that elementary particles have an intrinsic property called *spin*. Spin can be thought of as an intrinsic angular momentum, i.e. the electron is always spinning around an axis, that has to be included with orbital angular momentum in determining the motion of an elementary particle. Spin is as important as mass or charge in characterizing a particle at the quantum level. For example, the electrons, protons and neutrons introduced above all have spin value of $\frac{1}{2}$. All elementary particles with half-integer spin are known as *fermions*. The *Pauli Exclusion Principle* is another fundamental result in quantum mechanics and states that no two identical spin-$\frac{1}{2}$ particles can be in the same quantum state. This has the consequence that atomic electrons order themselves in shells, which is the basis of the Periodic Table of the elements. Further, it means that protons and neutrons also organize themselves into shells at the nuclear level. This can explain many of the important properties of nuclei.

In quantum mechanics, the force between two particles takes place via exchange of particles with integer spin, known as *bosons*.

Fig. 1.3: The German physicist Werner Heisenberg, who originated a formulation of quantum mechanics complementary to that of Schrödinger. In particular, he introduced matrix mathematics to mechanics.

Each force is mediated by a different boson. The boson exchange is not detectable directly and takes place in a very short amount of time. The range and strength of the force is governed by the mass of the boson exchanged. A large boson mass implies a short-ranged, weaker force than that for the case of a lighter boson mass.

1.1.3 *Standard Model of physics*

Over the last 100 years or so, physicists have developed an increasingly sophisticated and successful description of the world of atoms using quantum mechanics. This view is now encoded in the Standard Model, which we describe here.

Physicists understand the natural world in terms of four forces:

1.1.3.1 *Gravity*

The *force of gravity* acts at long distance scales on large masses. It is responsible for objects falling down at the Earth's surface, explains the motion of the planets in the solar system, and is essential to understand the large-scale structure of the universe. Gravity is described by Einstein's theory of General Relativity, where the effects of gravity result in an influence on the curvature of four-dimensional space–time. In February 2016, physicists announced the discovery of gravitational waves, predicted by Einstein's theory. The gravitational waves detected were generated by the interaction of two so-called black holes (astrophysical objects whose gravity is so strong that light cannot escape) far away in space. This gives rise to a ripple in space–time that is detected simultaneously as a tiny motion (about 1,000 times smaller than the size of a proton) by sophisticated detectors in Louisiana and Washington State in the USA. At this time, the boson associated with a quantum theory of gravity is known as a graviton and is predicted to have spin 2.

1.1.3.2 *Electricity and magnetism*

The quantum theory of *electricity and magnetism* is the most precisely tested theory in all of physics. As discussed above, it introduces fundamental new concepts which have become essential to our understanding of the atom and its constituents. Further, the electron has an anti-particle called the positron, with positive charge. The electrons and positrons interact by exchanging the quantum of light called the photon. The photon is the boson of *quantum electrodynamics* (QED) and has spin value 1. Positrons and electrons can annihilate to produce photons. In this picture, the vacuum is not empty but can fluctuate into electron–positron pairs. These effects have been measured and found to be in excellent agreement with predictions.

1.1.3.3 *Weak force*

The feebleness of the *weak force* results from the fact that it arises from the exchange of very heavy spin-1 particles, known as intermediate vector bosons. The weak force plays a key role in nuclear reactions, e.g. the generation of energy in stars like our Sun. In the 1980s, physicists produced a unified quantum theory of electromagnetism and the weak force. A prediction of this unified theory was the existence of the intermediate vector bosons, which were subsequently discovered. They come in three types, the uncharged Z^0, and the charged W^+ and W^-. They have masses about 90 times the mass of a proton. In addition to the electron, there are two heavier, point-like partners, called the muon (about 200 times the electron mass) and the tau (about twice the proton mass). Each of the electron, muon, and tau has an associated neutrino, which is known to have a small mass. All these point-like particles that do not take part in the strong force but rather participate in the weak interaction are known as *leptons*.

1.1.3.4 *Strong force*

The *strong or nuclear force* is the force that binds the protons, neutrons, and nuclei of atoms. It fuels the stars and explains the origin of the elements in our universe. It is the force that is harnessed in nuclear power reactors to produce electricity and is widely used in nuclear medicine. The fundamental theory involves interactions between six point-like quarks (given the names *up, down, strange, charm, bottom* and *top*) where spin-1 massless gluons are exchanged. The quark masses vary from a few times the electron mass (*up* and *down* quarks) to 180 times the proton mass (*top* quark). The fundamental theory of the strong force is known as *quantum chromodynamics* (QCD). The proton is understood to be composed of predominantly light *up* and *down* quarks moving at very high speeds and undergoing strong interactions that produce quark–anti-quark pairs and gluons. The quarks and gluons are not visible, only composite systems, like the protons and neutrons, with certain configurations.

It is instructive to consider the different strengths of the four forces. In general, they depend on the energy scale we are considering but, at the scale of the world we live in, the strengths of the strong, weak, and electromagnetic forces are in the ratio $\frac{1}{3}:\frac{1}{29}:\frac{1}{137}$. By comparison, the strength of the gravitational force between two electrons is 45 orders of magnitude smaller! Gravity is a negligible force in the sub-atomic world. Theoretical calculations are usually

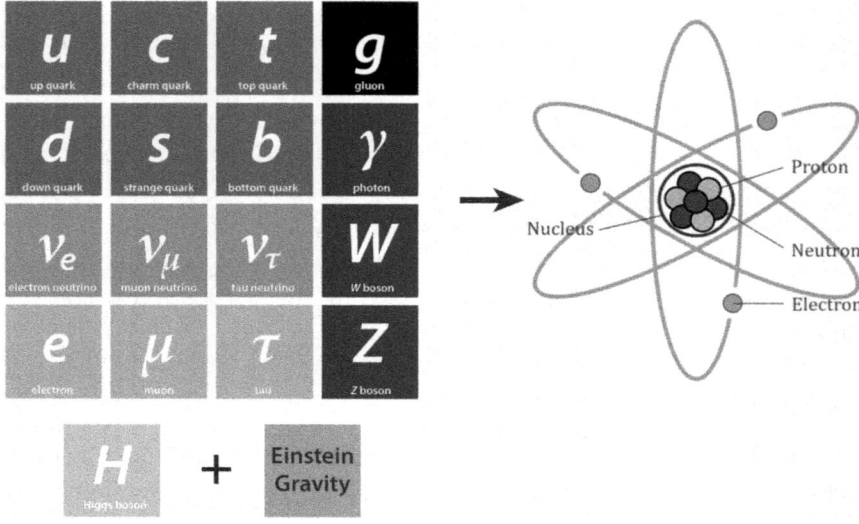

Fig. 1.4: Basic elements of the Standard Model. Shown are the quarks and leptons, which as far as we know, have no internal structure and are point-like, together with the field quanta of the electromagnetic, strong, and weak forces. The recently discovered Higgs boson H explains the masses of the particles. Gravitation has not yet been incorporated into the Standard Model.

carried out as a convergent series in powers of the strength of the force. A precise calculation is easier to achieve if the strength of the force is smaller. Thus, it is easy to see why quantum electrodynamics is the most precisely tested theory and why the strong force is difficult to calculate with precision at the typical energies of the universe around us.

The quantum description of the electroweak and strong forces comprises the Standard Model of physics. Together with Einstein's theory of gravity, it can explain all measurements in the laboratory. Figure 1.4 summarizes the elements of the Standard Model. Further, the Higgs particle was predicted to explain the masses of the quarks and leptons. In 2012, physicists using the Large Hadron Collider at CERN announced the discovery of the Higgs particle, which was found to have a mass about 125 times that of the proton. Of the 110 Nobel Physics Prizes awarded to date, more than 40 have been awarded for work leading to the development of the Standard Model, summarized in Fig. 1.4. It is difficult to identify another human achievement that is comparable. The Standard Model underpins our modern human civilization, e.g. cell phones, the Global Positioning System, the internet, and nuclear medicine, power and weapons.

It is important to point out that the Standard Model involves many parameters put in by hand. Further, it does not explain important phenomena observed in our universe, e.g. the existence of dark matter and dark energy, as well as the preponderance of matter over anti-matter. It is clear that the Standard Model is incomplete and extensions to it are widely sought.

1.2 Spin: Provider of Order and Structure to the Universe

We have seen that the spin of the elementary particles of the Standard Model is fundamental to their role and to the observed periodic/shell structure of the small part of the universe we inhabit, both at the atomic and nuclear level. In Appendix C, we provide a short history of spin.

The concept of spin originated in 1925 with young graduate students named Samuel Goudsmit and George Uhlenbeck in the group of Prof. Paul Ehrenfest at the University of Leiden in the Netherlands (see Fig. 1.5). Against the criticisms of renowned physicists, but with the firm support of Ehrenfest and Niels Bohr, Goudsmit and Uhlenbeck postulated that the electron had an intrinsic orbital angular momentum, i.e. spin of $\frac{1}{2}$ to explain a mysterious doubling observed in the spectral lines of atoms. Ehrenfest is quoted by Uhlenbeck as saying: "This is a good idea. Your idea may be wrong, but since both of you are so young without any reputation, you would not lose anything by making a stupid mistake." Within a few years, it had been established that spin was essential if one combined quantum mechanics and Einstein's theory of relativity. Fully relativistic theories of the electron and a non-relativistic theory of the protons and neutrons in the nucleus were developed by 1950 (Fig. 1.6).

Fig. 1.5:   The Swedish theoretical physicist Oskar Klein on the left with George Uhlenbeck and Samuel Goudsmit, the graduate students of Paul Ehrenfest at the University of Leiden, who introduced spin in 1925. The photo was taken in summer 1926. Photograph by H. Knauss, University of Leiden (Holland), courtesy AIP Emilio Segrè Visual Archives, Physics Today Collection.

However, in the laboratory at the middle of the 20th century, spin was almost unknown, as experimental techniques had not yet been developed. The next 30 years witnessed a remarkable development in experimental spin techniques. To understand this, we need to introduce the term *spin polarization*, which is the degree to which the spins of a collection of elementary particles are aligned along a given direction. A polarization of zero results from a completely random orientation of the spins. A polarization of 100% means that all the spins are aligned in the same direction. For effective experimental studies of spin, beams and targets with high spin polarization are necessary, as the observables requiring measurement are often small. The latter half of the 20th century witnessed great progress in the development of spin polarized beams of ions and electrons as well as targets. In the 1950s, the first polarized proton

Fig. 1.6:   Picture of the famous physicists Wolfgang Pauli (originator of the Pauli Exclusion Principle) on the left and Niels Bohr (one of the originators of quantum mechanics) on the right studying a child's spinning top. The picture was taken at the opening of the new institute of physics at the University of Lund on May 31, 1951. Credit: Photograph by Erik Gustafson, courtesy AIP Emilio Segrè Visual Archives, Margrethe Bohr Collection.

beams were developed. Early experiments demonstrated that spin played a crucial role in the forces between protons and neutrons which give rise to bound nuclei and that parity conservation (two physical systems, one of which is a mirror image of the other, must behave identically) was violated by the weak force. Polarized electron beams only became available for the first time in the 1970s. In 1978, parity violation was observed in scattering of high-energy, polarized electrons. This demonstrated the existence of the Z^0 intermediate vector boson and validated the electroweak theory, a cornerstone of the Standard Model.

Spin polarized targets of hydrogen were developed using low temperature techniques and high magnetic fields starting in the 1960s. High spin polarizations were achieved but the polarized nuclei were often contained in chemical compounds with a large amount of extraneous material. This had the consequence that the spin effects were diluted and difficult to measure precisely. However, in the 1970s, pioneering experiments using polarized electrons scattering from polarized protons were carried out at Stanford, CA. These experiments allowed a determination of the contribution of the quarks to the spin of the proton and were consistent with expectations. Thus, further experiments were not approved and the experimentalists moved to CERN, Geneva, Switzerland where such experiments could be pursued at higher energies using muon beams. The CERN experiments were carried out by the European Muon Collaboration (EMC).

While spin is a fundamental aspect of the quantum world, it is important to point out that it is used daily worldwide as a non-invasive, medical diagnostic tool. Magnetic resonance imaging (MRI) relies on reversing the spins of protons in a very high magnetic field. The resulting signals can be used to form an image which allows physicians to evaluate various parts of the body and determine the presence of certain diseases. Further, the technique has been expanded to include also polarized noble gases.

1.3   Modern View of the Proton

We conclude this chapter with a brief description of the structure and properties of the proton as viewed through the Standard Model. The proton is composed of two *up* quarks and a *down* quark. These three light, so-called *valence quarks* move at relativistic speeds and experience very strong forces due to exchange of gluons. These interactions produce further quark–anti-quark pairs (known as *sea quarks*) as well as gluons. See Fig. 1.7 for a schematic illustration of the proton's structure. In this picture, the properties and structure of the proton arise from these complicated interactions among its fast-moving constituents.

While a theoretical framework exists in the Standard Model to calculate the structure of the proton, the magnitude of the strong force is large, as described above, and so precise calculations at the typical energies in Nature are impossible with present techniques. Thus, simulations using large computers are the most direct means

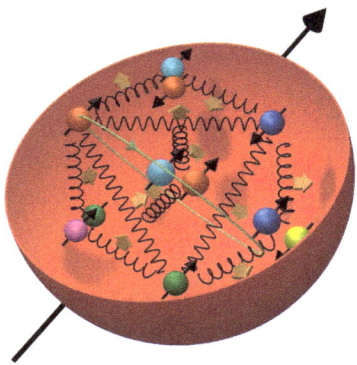

Fig. 1.7: Schematic picture of the proton's structure showing three valence quarks, quark-antiquark pairs and gluons. (*Source:* Julian Rith).

to do this at present. This technique is known as *lattice QCD*, and an essential element of the calculation is to discretize time and space on a lattice. Unfortunately, due to the huge amounts of computer cycles necessary for these calculations, most of the simulations to date have significant approximations. As more powerful computers become available, it is expected that these simulation techniques will increase in accuracy.

It is known that the gluons play a crucial role in proton structure. They account for about half of the momentum of the proton's constituents but because they are uncharged electrically, they are largely hidden from electrically charged probes such as the electron or proton. Gluons can split and recombine and this can have a sizable effect. Understanding their contribution to the proton's structure and properties is a major avenue for future research.

Another important subject for study is the process by which a struck quark or gluon in a high-energy collision materializes as a constituent in an experimentally detectable particle. This process, called *hadronization* connects the hidden world of the fundamental quarks and gluons of the Standard Model with the world of measurement. It relates to one of the major unsolved problems of physics, namely why are the quarks of QCD not detectable as free particles? This is known as *confinement*. To date, there has been no proof of confinement from the Standard Model.

In summary, while we have a fundamental theoretical framework from the Standard Model to understand the strong force, we are far from having a satisfactory understanding of the proton, neutron, and nuclei in terms of the quarks and gluons of QCD. It is a major thrust of worldwide physics research to achieve this and the HERMES experiment has provided important insight which has guided the path forward.

1.4   Experimental Methods to Study the Fundamental Structure of Matter

Much of our knowledge about the sub-atomic world, i.e. objects below the size of an atom, results from scattering experiments using charged and neutral particles, like electrons or ions. A beam of particles a impinges on a target of particles b; a scattered particle c is detected at angle θ with respect to the incident beam direction and an unobserved remnant X remains. We can denote this process by

$$a + b \rightarrow c + X.$$

In the years 1908–1913, a landmark series of experiments carried out by Hans Geiger and Ernest Marsden, under the direction of Ernest Rutherford (Fig. 1.8) at the Physical Laboratories of the University of Manchester, discovered that every atom has a nucleus where all of its positive charge and most of its mass reside. They determined this by careful measurements and insightful analysis of the scattering experiment $\alpha + \text{Au} \rightarrow \alpha + X$. A collimated beam of α-particles (nuclei of the ^{4}He atom) from a radioactive source was directed onto a thin gold (Au) foil. The scattered α-particles were observed under a microscope by their scintillations in a crystal. The experiments showed that most of the α-particles were transmitted, and a few deflected by the target, some of them to larger angles.

The analysis by Rutherford led to a novel model of the gross structure of atoms, a small massive nucleus of positive charge, surrounded by an extended electron cloud, the so-called Rutherford model of atoms (Fig. 1.9). Previous to the Manchester experiments, the popular theory of atomic structure was the *plum pudding model*. This model, developed by J.J. Thomson, postulated that the atom

Fig. 1.8:   The New Zealand-born experimental physicist Ernest Rutherford who led the experiment that discovered the atomic nucleus at the University of Manchester. (*Source*: Nobel Foundation Archive).

was a sphere of positive charge throughout which the electrons were distributed, like raisins in a Christmas pudding. This model was based entirely on classical Newtonian physics and both protons and neutrons were unknown then. When Geiger and Marsden shot alpha particles at their metal foil, they noticed only a tiny fraction of the alpha particles were deflected by more than 90°. Most flew straight through the foil. This suggested that those tiny spheres of intense positive charge were separated by vast gulfs of empty space. Most particles passed through the empty space and experienced negligible deviation, while a handful passed close to the nuclei of the atoms and were deflected through large angles. Rutherford thus rejected Thomson's model of the atom, and instead proposed a model where the atom consisted of mostly empty space, with all of its positive charge concentrated in its center in a very tiny volume, surrounded by a cloud of electrons.

When Geiger reported to Rutherford that he had spotted alpha particles being strongly deflected, Rutherford was astounded. In a lecture Rutherford delivered at Cambridge University, he said: "It was quite the most incredible event that has ever happened to me in my life. It was almost as incredible as if you fired a 15-inch shell at a piece of tissue paper and it came back and hit you. On consideration, I realized that this scattering backward must be the result of a single collision, and when I made calculations I saw that it was impossible to get anything of that order of magnitude unless you took a system in which the greater part of the mass of the atom was concentrated in a minute nucleus. It was then that I had the idea of an atom with a minute massive centre, carrying a charge."

1.5 Experimental Study of the Quark and Gluon Structure of Matter

The Manchester experiments in 1908–1913 profoundly influenced the direction of sub-atomic physics experiments since that time. Study of the fragments resulting from collisions between energetic beams and targets has been the principal means to elucidate subatomic structure up to the present day. The most recent step in this energy race resulted in the Large Hadron Collider (LHC) at CERN with colliding proton beams of up to 7 TeV = 7,000 GeV in energy (see Appendix A for an explanation of these units). This has driven development of novel sources, accelerators, targets, and detectors over more than a century. Consequently, we have probed distance scales from 10^{-10} m in Rutherford's time to smaller than 10^{-20} m at the LHC.

Because electrons do not have any internal structure, they can be used as a precise probe of the more complicated nucleons and nuclei. A scattered electron creates a virtual photon to see inside the nucleon; the photon energy (technically the square root of Q^2, its total momentum squared) determines its resolving power, see Fig. 1.9. Groundbreaking experiments led by Robert Hofstadter at Stanford in the 1950s, recognized with the 1961 Nobel Prize, directly confirmed that nucleons are not elementary; rather, the charge distribution of the proton has a size of order 10^{-15} m.

To reveal the substructure of nucleons, higher spatial resolution is needed, which requires using electron beams of significantly higher

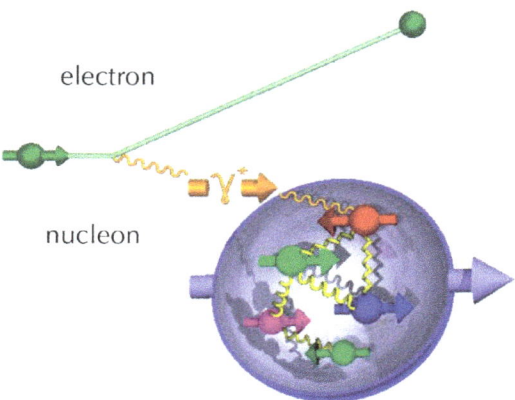

Fig. 1.9: An electron scattering from a proton exchanges a virtual photon with the charged constituents. (*Source:* U. Elschenbroich/HERMES).

Fig. 1.10: Richard Taylor, Henry Kendall and Jerome Friedman, the leaders of the MIT-SLAC experiment that experimentally discovered the point-like quarks in electron scattering from the proton in 1967. Courtesy of AIP Emilio Segrè Visual Archives.

energy. Of particular relevance to our story is the MIT-SLAC experiment that took place at the Stanford Linear Accelerator Center (SLAC) in the late 1960s. In 1967, the newly built 2-mile long electron accelerator at the SLAC enabled a new kind of electron-scattering experiment, known as deep-inelastic scattering (DIS), in which the energy is large enough to violently smash the proton target. A group of physicists, led by Henry Kendall and Jerome Friedman from MIT and Richard Taylor from SLAC (Fig. 1.10), measured the scattered electron rate in kinematics where the incident electron lost most of its energy. They observed that the rate in this deep inelastic regime varied slowly with momentum transfer. Analogous to Rutherford's analysis in Manchester, James Bjorken and Richard Feynman deduced that this unexpected large rate arose from scattering from point-like, fractionally charged constituents of the proton, so-called *partons*. Bjorken and Feynman's parton model allowed determination of the quark momentum distributions only in a special boosted reference frame. In this frame, due to relativistic effects, the proton was Lorentz contracted and the interaction time of the virtual photon with the proton's charged constituents was dilated (Fig. 1.9).

The whole picture came together when partons were identified as fractionally charged quarks. The interactions of the quarks, mediated by gluons, each of which carry a color (literally a type of internal charge analogous to electric charge) are described by the theory of QCD, developed in the 1970s. One central feature of QCD is "confinement", which is the locking together of quarks in hadrons. Unlike the electromagnetic force, the color force increases in strength as the distance between quarks or quarks and gluons increases, thus explaining why quarks and gluons do not exist as free particles, and why studying them inside nucleons and nuclei with electrons has been and will continue to be valuable to advancing scientific understanding. Nobel Prizes were awarded to Jerome Friedman, Henry Kendall, and Richard Taylor in 1990 for the SLAC experiments, and to David Gross, David Politzer, and Frank Wilczek in 2004 for their insights into how the color force works.

At high energies, electron scattering from the proton is analyzed in the boosted frame in terms of two experimentally measurable quantities: Q^2, the square of the momentum of the exchanged photon, and x the fraction of the proton's momentum carried by the struck quark. Measurements of the DIS rate for a given incident electron energy and in a detector at a given scattering angle provide a snapshot of the moving, fluctuating charged constituents of the proton. These snapshots as a function of Q^2 and x can be used to provide a visual picture of the proton's quark and gluon structure. In analogy with a camera, these snapshots are characterized by a focus with spatial resolution ($\sim \frac{1}{Q}$) and shutter speed ($\sim \frac{1}{x}$).

At small spatial resolution, corresponding to large Q^2, the photon can resolve new phenomena, such as the possibility that a quark radiates or absorbs a gluon or that gluons produce quark–anti-quark pairs. The quantity x is a measure of the energy of the exchanged photon, where large energy corresponds to small x. The physical meaning of x is most transparent in the picture in which the photon interacts with a single quark in the target, and x determines the fraction of the total momentum of the colliding nucleon carried by that constituent. DIS experiments have established a basic picture of the nucleon in which, at low resolution, the nucleon is composed of three valence quarks, each with x approximately $\frac{1}{3}$. With increasing resolution, additional sea quarks and gluons become visible, and these extra constituents dominate the small x regime.

At lower energies, electron accelerators were built in the 1970s to continue these studies on nuclei at laboratories around the world, including at Saclay in France, the Massachusetts Institute of Technology (MIT) Bates Lab in the United States, the National Institute for Subatomic Physics in the Netherlands, and at the universities of Bonn and Mainz in Germany. These accelerators revealed more details about the structure of the nucleon and the behavior of nucleons within the nucleus, and their development led to improved accelerator technology as well.

High-energy electron scattering experiments have continued at JLab, in the United States, from 1995 to the present, and at HERA in Hamburg, Germany, from 1991 to 2007. JLab, building on earlier work at SLAC, has pioneered new accelerator technology, including sources and acceleration of polarized electron beams, superconducting accelerators, higher-intensity beams, and more sophisticated detectors and analyses. HERA pioneered high-energy collisions of longitudinally polarized electron and positron beams with unpolarized proton beams and the H1 and ZEUS experiments contributed enormously to our understanding of the quark and gluon structure of proton, as described in Section 6.2. The HERA Measurement of Nucleon Spin (HERMES) fixed-target experiment, the focus of this book, pioneered the measurement of pure lepton-hadron scattering, without the complication of scattering from non-target material or from the end-windows, which is present in conventional experiments. HERA experiments revealed the great abundance of gluons within neutrons and protons.

In closing, it is important to understand how to make high-energy electron scattering sensitive to the spin of the constituents of the proton (quarks and indirectly gluons). This is made possible by ensuring that the spins of the incident electron beam and the proton target are not randomly oriented but have a preferred direction, e.g. along the incident beam direction, known as longitudinal. We say that the beam and/or target are *polarized longitudinally*. Then the DIS spin asymmetry A of scattered electrons for parallel and anti-parallel target spins characterizes the spin dependence of the measured process. Here the asymmetry A is defined by

$$A \equiv \frac{\uparrow\uparrow - \uparrow\downarrow}{\uparrow\uparrow + \uparrow\downarrow},$$

where the rates $\uparrow\uparrow$ and $\uparrow\downarrow$ correspond to the DIS rates for beam and target spins parallel and anti-parallel, respectively. Note that A is the same whether the beam or target spin is reversed so this is a good cross check on the measurement ($-1 \leq A \leq 1$). Measurement of this proton spin asymmetry enables the extraction of quark polarizations, i.e. the degree to which the quark spins are directed along the proton spin. Another asymmetry can be constructed from scattering when the proton target spin is polarized transverse to the beam, and the azimuthal distribution of leading hadrons from the debris of the target nucleon is measured with respect to the proton spin. It is called a Single Spin Asymmetry (SSA) and reflects the transverse polarization of the quarks in the proton. HERMES pioneered this type of measurement. Both longitudinal and transverse types of asymmetries have been measured. The question remains how the spins of the colliding particles can be prepared such that they fulfill the experimental conditions, i.e. a longitudinally polarized electron beam at the target position, and a thin (gas) target with polarization along or transverse to the beam. These are described in Chapter 3.

Chapter 2

The Proton's Spin

2.1 Introduction

In the early 20th century, physicists developed a new mechanics, so-called *quantum mechanics* to describe the subatomic world. Here, observables like energy, took discontinuous values, in stark contrast to classical mechanics, where they were continuous. Albert Einstein had explained the photoelectric effect by postulating that light was quantized in particles called *photons* with energy $E = h\nu$, where ν is the frequency of the light and h, known as Planck's constant, characterizes the quantum world. Max Planck had derived a universal law for radiation produced by atoms in thermal equilibrium with his constant h again playing a central role. See Appendix A for more details. As we shall see in this chapter, Niels Bohr derived a quantum theory of the motion of orbital electrons around the nucleus of the atom where angular momentum was quantized in units of $h/2\pi$.

2.2 The Origins of Spin

Since the mid-19th century, it was known that each element, when excited to a sufficiently high temperature, emitted a visible series of discrete *spectral lines* at specific wavelengths which were unique to that element. For example, the element *helium* (named for the Greek god of the Sun, *Helios*) was first discovered not on Earth but in the spectral lines observed in sunlight during a solar eclipse in 1868. Simple formulae were developed to describe the wavelengths of the visible lines of hydrogen. Subsequently, it was discovered that these spectral lines could be influenced by a strong magnetic field. The number of lines could be doubled mysteriously, which was termed *duplexity*.

By the early 20th century, it became evident that atoms and molecules with even numbers of electrons are more chemically stable than those with odd numbers of electrons. For example, it was realized that the atom tends to hold an even number of electrons in the shell which surrounds the nucleus. In 1919, it was suggested that the periodic table of the elements could be explained if the electrons in an atom were connected or clustered in some manner.

About 1920, Niels Bohr developed the Bohr theory of the atom, in analogy with the motion of planetary orbits around the Sun. In Bohr's theory, the electrons circled the nucleus in stationary orbits subject to the constraint that the orbital angular momentum is *quantized*, i.e. is a multiple of a fundamental constant, see Fig. 2.1. When electrons jump from one orbit to another they gain or lose energy by specific amounts. Thus, the visible spectral lines are the result of electrons losing energy when transitioning between orbits. The Bohr theory predicted the simple formulae that had been developed previously from data. This was a major milestone in explaining the spectral lines in terms of a quantum theory of the atom. However, as noted above, the number of lines observed experimentally was mysteriously double what was predicted by the Bohr theory.

In 1922, Stern and Gerlach carried out their famous experiment of passing a beam of silver atoms through an inhomogeneous magnetic field and observing a separation into two beams (see Fig. 2.2). At the time, the experiment was interpreted as a crucial validation of the Bohr theory over the classical theory of the atom. It showed clearly that spatial quantization exists, a phenomenon that can be accommodated only within a quantum mechanical theory. In 1925, the Pauli Exclusion Principle was formulated as: *no two electrons can have identical quantum numbers.*

Most significantly for this discussion, and as we have mentioned in Chapter 1, also in the year 1925, Leiden graduate students Uhlenbeck and Goudsmit first hypothesized intrinsic spin as a property of the electron. Primarily, they were motivated by the desire to explain the doubling of states. However, others had considered

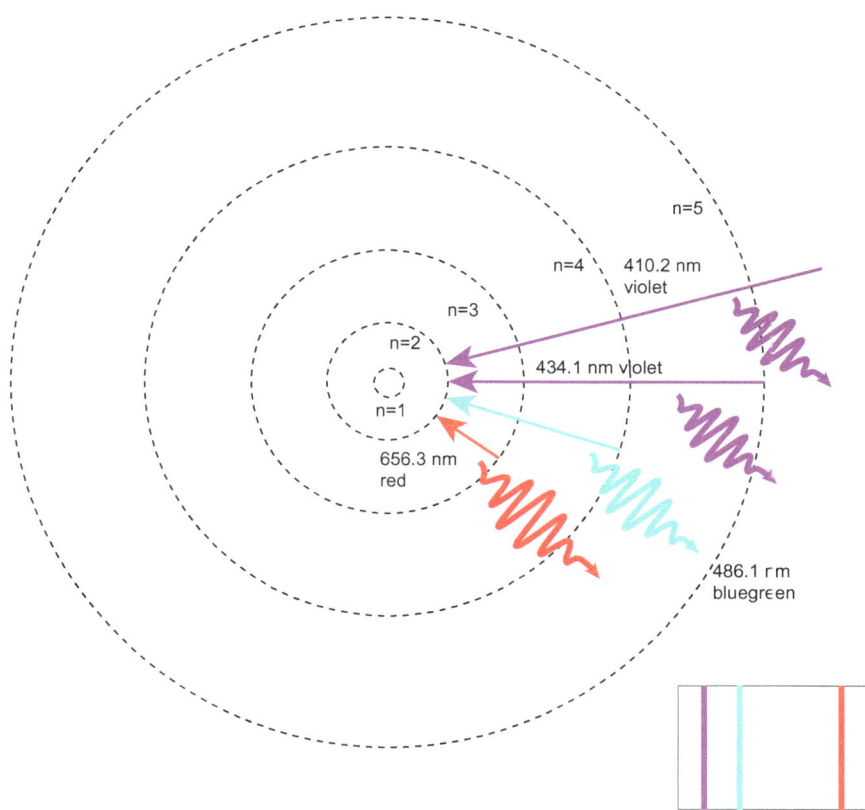

Fig. 2.1: Schematic layout of the hydrogen atom as explained by the Bohr theory. Each level is characterized by an integer n and the light emitted in transition from level n_1 to level n_2 has an energy ΔE given by the Balmer formula. The frequency of the light is given by $\Delta E = h\nu$ where h is Planck's constant. The spectral lines of hydrogen in the visible spectrum are shown in the lower right of the figure and the corresponding transitions are indicated in the atom.

this possibility and rejected it on fundamental grounds. At that time, the electron was believed to have a finite size, estimated to be about the currently known radius of the proton. If the electron was spinning, then the speed at the surface of the electron exceeded by about an order of magnitude the speed of light. This was believed to be a violation of Einstein's Theory of Special Relativity which was completely accepted by the mid-1920s. In particular, the famous Dutch physicist Lorentz strongly objected on these grounds and made

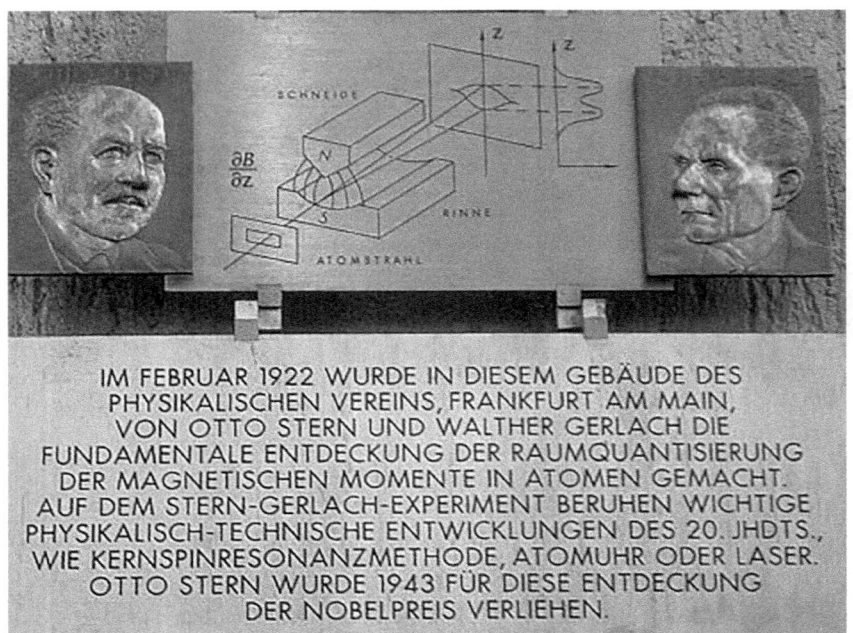

Fig. 2.2:   Plaque at the University of Frankfurt commemorating the Stern–Gerlach experiment. German text translated: *In February 1922 in this building of the Physikalischer Verein, Frankfurt am Main, Otto Stern and Walther Gerlach made the fundamental discovery of the spatial quantization of magnetic moments in atoms. Important physical-technical developments of the 20th century, such as the nuclear spin resonance method, atomic clock or laser, are based on the Stern–Gerlach experiment. Otto Stern was awarded the Nobel Prize for Physics in 1943 for this discovery.* It should be noted, that the 1943 Nobel Prize in Physics was awarded not for the Stern–Gerlach experiment, but "for the invention of the molecular beam method and the measurement of the g-factor of the proton".[1]

this clear to both Uhlenbeck and Goudsmit. However, their advisor Paul Ehrenfest had already submitted their paper! In their *Nature* letter they write: *It seems that the introduction of the concept of the spinning electron makes it possible throughout to maintain the principle of the successive building up of atoms utilized by Bohr in his general discussion of the relations between spectra, and the natural system of the elements. Above all, it may be possible to account for the important results arrived at by Pauli without having to assume an unmechanical duality in the binding of the electrons.* In the succeeding letter in the same journal, Bohr fully agreed.

Fig. 2.3: The British theoretical physicist Paul Dirac, who derived the relativistic quantum mechanical equation for the electron that bears his name. The Dirac equation provides theoretical motivation for the antiparticle and for the spin of the electron. (*Source*: Florida State University Libraries' Special Collections and Archives Division).

In 1928, the British physicist Paul Dirac, motivated by the desire to construct a quantum theory that was consistent with Einstein's theory of relativity, developed his elegant equation for spin-$\frac{1}{2}$ particles (Fig. 2.3). In this formulation, the solutions have four components which are interpreted as positive and negative energy states of spin $\pm\frac{1}{2}$ each. Dirac predicted the existence of the antiparticle of the electron, called the *positron*, and the theory became the basis for the most precisely tested theory in physics, Quantum Electrodynamics. By the end of the 1920s, physicists had developed a fundamental understanding of the essential role of electron spin in explaining the electronic structure of the atom. There exist excellent, personal, historical accounts by Dirac,[2] Uhlenbeck,[3] and Goudsmit[4] of this period. A comprehensive biography of Paul Dirac is available.[5]

By the middle of the 20th century, the intrinsic spin of subatomic particles was a cornerstone of the physicist's theoretical understanding of the fundamental structure of matter at both the electronic and nuclear levels. However, spin as an experimental tool became a reality only in the second half of the 20th century. See Appendix C for details. In the 1950s, a number of seminal experiments were carried out using spin. In 1956, the theorists Lee and Yang pointed out that parity should be violated in the weak interaction. Parity inversion is the effect of looking at the mirror image and it was a surprise that the weak interaction violates this symmetry. Shortly afterwards, in 1956, Chien-Shiung Wu and collaborators observed parity violation in the aligned nucleus ^{56}Co. In 1958, it was shown experimentally using polarization techniques that the neutrino, one of the fundamental particles of the Standard Model, has negative *helicity*, which is its angular momentum in the direction of its momentum.

2.3 Quarks Arrive

With the availability of new, higher energy particle accelerators in the 1950s, physicists studying the fundamental structure of matter discovered a large number of new particles. These were all bound by the strong force and they were so plentiful that they were labeled the "particle zoo". In reaction, Wolfgang Pauli is said to have exclaimed "Had I foreseen that, I would have gone into botany." There was a clear desire to organize the "zoo" in terms of a theory of a few simple constituents. This was established in 1964 by Gell-Mann, and independently by Zweig, with the conjecture of elementary spin-$\frac{1}{2}$ constituents called *quarks*. In the original theory, there were three types or *flavors* of quarks: the *up* quark with electric charge $+\frac{2}{3}e$; the *down* quark with electric charge $-\frac{1}{3}e$ and the *strange* quark with electric charge $-\frac{1}{3}e$, where e is the charge of the proton. The *strange* quarks were so-called because the particles containing them were surprisingly long-lived. This was understood later to be due to the fact that particles which contained strange quarks decayed via the weak force. The proton could be considered as a composed of two *up* quarks and a *down* quark while the electrically uncharged neutron was formed from one *up* quark and two *down* quarks. Thus, the hundreds of strongly interacting particles in the "particle zoo" could be understood as composites of the elementary quarks. However, it was unclear whether the quarks were simply an abstract concept to facilitate in organizing the "zoo" or whether they were real constituents of the proton.

This all changed with the electron–proton scattering experiments at the new high-energy, two-mile long Stanford Linear Accelerator Center (SLAC) in the late 1960s. More than 50 years later, the SLAC linear accelerator is still the longest linear accelerator in the world. Beginning in 1966, SLAC made available high-energy electron beams that could probe the proton with an unprecedented fine spatial resolution. The initial experiments focused on detecting only the scattered electron. In particular, an MIT-SLAC collaboration concentrated on detecting scattered electrons that had lost much of their initial energy. They discovered a region where the effects of proton structure on the scattering were independent of the spatial resolution, so-called *deep inelastic scattering* (DIS). This was interpreted by Bjorken, Feynman and others that the proton was composed of point-like constituents. A key determination was made experimentally that these constituents had spin-$\frac{1}{2}$ which led to their identification as the same quarks which Gell-Mann and Zweig had invented to organize the "zoo" (Fig. 2.4). Thus, we knew that

Fig. 2.4: Theoretical physicist James Bjorken from SLAC and Fermilab. Bjorken's insights were key to interpreting the high-energy electron scattering measurements from the proton at SLAC in the late 1960s.

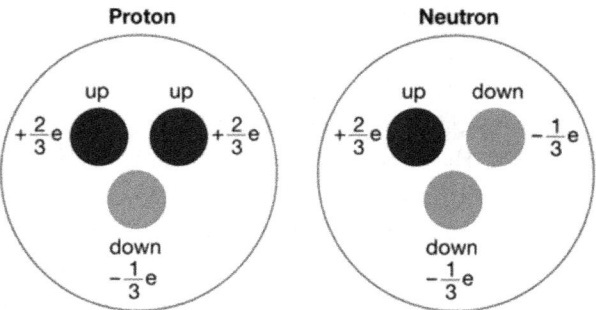

Fig. 2.5:   Quark model schematic picture of the proton and neutron showing the *up* and *down* quarks.

quarks were real and for the first time a model of the proton in terms of fundamental, point-like constituents, the quarks, could be constructed. Figure 2.5 shows schematically the quark structure of the proton and neutron.

The quark model gives a satisfactory explanation of the observed masses of the many particles in the "zoo" and even predicted some that were only later observed experimentally. It can straightforwardly explain the origin of the proton spin as arising from the spins of the quarks together with their orbital angular momentum. The quarks move with relativistic speeds so, in the quark model, we expect that only about 60% of the proton's spin results from the intrinsic quark spins. More sophisticated versions of the simple quark model described here have been developed. For example, the role of the strange quarks has been considered. John Ellis and Robert Jaffe developed an important sum rule for each of the proton and neutron, assuming that the strange quarks carry no spin. A more fundamental sum rule had been derived by James Bjorken in 1966 that required the difference of measurements on the proton and neutron. Verification of these sum rules would require high-energy electron–proton measurements of the type pioneered at SLAC but with both the electron beam and the target polarized (Fig. 2.6).

Fig. 2.6: MIT theoretical particle physicist Robert Jaffe who has been at the forefront of understanding the origin of the proton's spin for several decades.

2.4 Challenges to Experimental Study of Spin

By the middle of the 20th century, the intrinsic spin of subatomic particles was a cornerstone of the physicist's theoretical understanding of the fundamental structure of matter. As described earlier, both the electronic shell structure of atoms and the *magic numbers* of nuclei (specific numbers of protons and neutrons corresponding to closed shells) were understood as direct consequences of the Pauli Exclusion Principle to the electrons and nucleons, respectively. However, in 1950, polarized beams and targets still did not exist. Spin only became an experimental tool in the period 1950–1975: see Appendix C and references therein. Here, we provide the reader with a brief overview of the capabilities in polarized electron beams and polarized proton targets before HERMES.

2.4.1   *Polarized electron sources*

Although techniques existed to polarize atoms by the 1950s, sources of polarized electrons came much later. In 1963, Vernon Hughes and colleagues began consideration of polarized electron sources at Yale. They pioneered the development of the PEGGY source in 1971–1974 which produced polarized electrons by taking a polarized lithium atom and detaching the polarized electron by means of a flash lamp. This PEGGY source produced highly polarized electrons of about 80% but with lower intensity by a factor of 400 compared to the unpolarized beam at SLAC beginning in summer 1974 (Fig. 2.7).

Also at SLAC, experiment E122, led by Charles Prescott, for the first time used a polarized electron source based on optical pumping of GaAs, invented by Pierce and Meier (ETH Zürich) in 1975. This technology dramatically enhanced the ability to carry out experiments using polarized electron beams. At that time, the electron beam polarization was limited to about 40%. In 1978, E122 announced the observation of parity violating electron scattering at SLAC for the first time. This validated the Weinberg–Salam Model

Fig. 2.7: Experimental physicist Vernon Hughes from Yale University who pioneered measurement of spin-dependent DIS at SLAC and at CERN.

and provided the first measurement of the neutral current coupling of the electron.

In 1963, Russian theorists A. A. Sokolov and I. M. Ternov predicted the process of *radiative polarization*, by which ultra-relativistic electrons acquire spin polarization when circulating in a storage ring. In 1968, Sokolov-Ternov self-polarization of electrons was first observed at the ACO storage ring at Orsay (Paris), and successfully verified in 1971 with the 625 MeV electron beam at the VEPP-2 storage ring, Novosibirsk, Russia (see also Section 3.7).

2.4.2 *Polarized proton targets*

The earliest attempts at making a polarized proton target used static methods, i.e. low temperatures and high magnetic fields to align the proton spins. This produced low polarizations. The big breakthrough was the development of *dynamic nuclear polarization* (DNP), which has been used for the majority of solid polarized targets used in nuclear and high-energy physics. This technique began in 1953, when Overhauser at Illinois proposed to transfer the polarization of conduction electrons in a metal to nuclei. Initially met with great skepticism, Overhauser's suggestion was experimentally verified later that year. Working independently, Jefferies and Abragam both suggested to dynamically polarize nuclei by coupling electron and nuclear spins. In 1962, Abragam, Borghini and co-workers built the first polarized proton target for the 20 MeV polarized proton beam at Saclay. Shortly thereafter, Chamberlain, Jefferies, and collaborators built a polarized proton target for 250 MeV pion scattering experiments at Berkeley. In both cases the average polarization of protons in target material was only about 20%, but steady improvement would be made in the following years, thanks to refinements in magnets and cryogenics. See Section C.3.3 for a more complete discussion of polarized proton targets.

A fundamental limitation of these targets is the presence of a large amount of extraneous material other than the polarized protons. In a scattering experiment, this dilutes the scattering asymmetry and gives rise to complicated corrections which must be introduced to extract the physics of interest. It is one of the great advantages of the polarized internal gas target technique that the target gas is the pure polarized atoms and the dilution and corrections vanish.

2.5 First Experiments to Study the Proton's Spin Structure

Following the discovery of quarks and the success of the quark model in explaining the masses of observed particles, it was desirable to next directly measure the polarization of the quarks in the proton. The obvious way to do this was to implement a polarized electron beam at SLAC and direct them on a polarized proton target, i.e. carry out the DIS measurements but now use spin. However, a major impediment to carrying out experiments requiring polarized electron beams was the fact that no source of polarized electrons existed in the 1960s. Atomic beam sources of polarized atoms had been developed in the 1950s.

Thus, in the 1970s, the SLAC experiments E80 and E130 measured spin-dependent DIS from the proton for the first time. The polarized proton target was built by a Yale–SLAC collaboration and used a solid butanol sample at low temperature in a high magnetic field. The electron beam produced significant radiation damage in the target during the running of the experiment which had to be addressed. Figure 2.8 shows the measurements of the proton spin asymmetry A_1 vs. x; A_1 is as defined for longitudinal beam and target polarizations in Section 1.5. It was found that the valence quarks in the proton, accessible at $x > 0.3$, were significantly polarized as expected. However, at low $x \sim 0.1$, the asymmetry was significantly lower. At SLAC, plans to construct the SLC were underway, so a subsequent proposal led by Hughes to probe the valence quark region in the neutron was turned down.

2.6 The Proton Spin Crisis

The groundbreaking experiments at SLAC were limited by both statistics and kinematics. The relatively low electron-beam energy

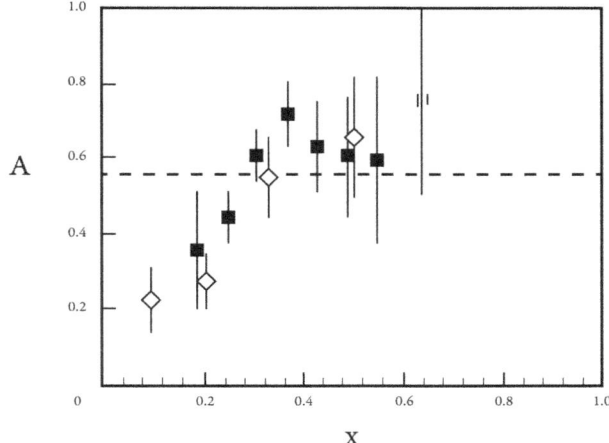

Fig. 2.8: Measured values of A_1 vs. Bjorken x obtained in SLAC E80 (open diamonds) and E130 (closed squares). The dashed horizontal line denotes the quark model expectation of $\frac{5}{9}$ for valence quarks, i.e. for $x > 0.3$.

at SLAC kept Q^2 well below the asymptotic regime, especially for x less than 0.1. Therefore, the evaluation of the sum rule depended on a speculative extrapolation to low x. The low-x behavior of the structure functions remains a matter of some controversy even to this day.

Next the experimental focus shifted to the Super Proton Synchrotron (SPS) at CERN, where the European Muon Collaboration, already famous for its work on deep inelastic muon scattering off unpolarized nuclear targets, now mounted a major effort to measure spin-dependent DIS from the proton. From the parity-violating decay of pions at SPS energies one could make a muon beam with energies up to 200 GeV and a natural longitudinal polarization of about 80%. At such high energy, the EMC collaboration was able to reach much higher Q^2 and lower x than the SLAC experiments had attained. In any case, SLAC had now shifted its attention to its linear-collider program. So, Vernon Hughes and his Yale group crossed the Atlantic ocean to join EMC.

The EMC collaboration reported its first results on spin-dependent DIS from the proton informally at several conferences early in 1987. The results were entirely unexpected: The combined CERN and SLAC data violated the Ellis–Jaffe Sum Rule for the proton. The new CERN data were consistent with the astonishing notion that none of the proton's spin is attributable to the intrinsic spins of its up, down or strange quarks!

Equivalently, the analysis of the EMC data indicated that a significant fraction of the proton spin is carried by strange quark–antiquark pairs. Although the EMC results were statistically consistent with the previous SLAC results and less than three standard deviations away from the theoretical expectation, they nonetheless ignited a wildfire among theorists and stimulated other experimental groups to extend these measurements.

The EMC results on proton spin in 1987 came as a great surprise. Why should the quark model that had worked so nicely fail so badly? Indeed, for some the result threatened to undermine quantum chromodynamics (QCD) — the theory put forward in the early 1970s that describes how the strong nuclear force acts between quarks. There were conjectures that the experiment was wrong, or even that QCD was wrong.

The seriousness of the situation in the late 1980s was summed up in the name many physicists gave to it: "spin crisis". Reluctant to ditch the quark model because of its substantial successes, researchers instead devoted their energies to finding alternative sources of the proton spin. There were several possibilities. It could have come from the angular momentum acquired by quarks and gluons — the particles that carry the strong nuclear force and glue the quarks together inside protons and neutrons — as they rotate about the protons spin axis. However, this orbital angular momentum is hard to measure. Many researchers instead pinned their hopes on another option: the spin of gluons.

2.7 Where Do the Proton and Neutron Really Get Their Spin?

By 1995, after several further rounds of experiments by EMC and other groups at CERN and Stanford, the result had shifted somewhat: Quark spins appeared to account for 20–30% of the spin of the proton or neutron. But the fact remained that much of the nucleon's spin must lie elsewhere. For sure, the relative orbital angular momentum of the quark and gluon constituents must play a role. However, accessing directly these contributions to the spin of the proton is fraught with challenges because of the highly relativistic, quantum-mechanical nature of the proton's structure. A surprising corollary is that the sea of quark–antiquark pairs that reside with the three constituent (valence) quarks is strongly polarized in the direction opposite to the nucleon's net spin. Furthermore, this sea appeared to contain a surprisingly large admixture of polarized strange quarks. This observation launched a program of parity violating electron scattering experiments over several decades, which thus far has not observed any sizable contributions from strange quarks to the proton's magnetism.

The so-called spin crisis has had important effects beyond simply confronting theorists with a particularly sharp challenge to their incomplete understanding of QCD. By validating the Bjorken sum rule it had in fact given us a striking confirmation of QCD. The spin crisis has also added renewed urgency to the quest for an understanding of nucleon structure, and it has given new impetus to the use of high-energy lepton probes (electrons, muons and neutrinos) of the fundamental quark and gluon structure of the nucleon. Since the mid-1990s, the scientific interest in understanding QCD using high-energy lepton scattering has grown significantly. As we shall see, the HERMES experiment played a major role in this evolution.

Chapter 3

HERMES: A New Way to Study the Origin of Proton and Neutron Spin

3.1 Overview

The HERMES experiment originated from two distinct sources: one in Europe led by Klaus Rith at the Max Planck Institute for Nuclear Physics in Heidelberg, Germany and the second in North America led by Richard Milner at the Kellogg Radiation Laboratory at Caltech, Pasadena, California. Strongly motivated by the unexpected results from the EMC measurements on proton spin described in Chapter 2, both the European and North American groups independently and simultaneously formulated letters of intent in 1988 to the DESY laboratory at the HERA electron storage ring to carry out measurements of spin-dependent electron-nucleon scattering (a) to verify the surprising EMC results and (b) to extend the EMC measurements in an effort to understand the origin of proton and neutron spin. The very different approaches of the EMC and HERMES experiments are discussed later in Section 3.3.3. In this chapter, the origins of the HERMES experiment, the unification into a single, potent transatlantic collaboration and the formulation of a detailed scientific proposal to mount a technically innovative, precision experiment at HERA are described.

It is not by accident that the two authors played an important role in founding HERMES, as they were leaders of the development of the key technology that was essential for HERMES, namely the spin-polarized gas targets of sufficient density and polarization for utilization in storage rings. A key innovation was proposed at the

Fig. 3.1:   Willy Haeberli, from the University of Wisconsin-Madison, one of the
fathers of the polarized hydrogen and deuterium target, with colleagues Tom Wise
(on right, Wisconsin) and Alex Nass (on left, Juelich) (*Source*: W. Haeberli).

1965 Karlsruhe meeting by our colleague Willy Haeberli from the
University of Wisconsin-Madison (Fig. 3.1). He proposed a storage
vessel with a teflon wall coating, in analogy to Norman Ramsey's
storage bulb of his hydrogen maser.[6] This was noted with interest at
that time, but there were no storage rings where one could have used
this technique effectively. Fifteen years later, Haeberli had assembled
a group of talented students and set up a test experiment at the
Madison tandem accelerator, based on used parts, in order to confirm
his 1965 proposal. At the 1980 Santa Fe conference, he reported that,
despite about 900 wall collisions, the polarization was practically
unchanged compared with a free atomic beam.[7] This result paved
the way towards pure gas targets of high polarization. Later it turned
out that, instead of a vessel, a T-shaped tubular structure, as shown
in Fig. 3.2 (see caption) gives the highest target density increase,
which can be up to 100 times higher than without the storage cell.

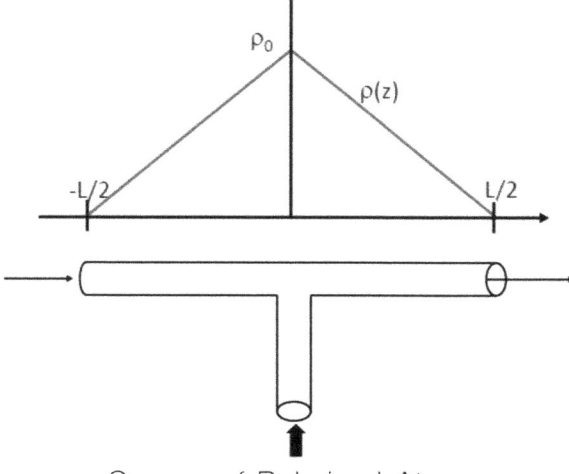

Source of Polarized Atoms

Fig. 3.2:   Schematic diagram illustrating the principle of the storage cell target. The beam traverses a straight tube of length L. Gas from the source of polarized atoms is injected into the center, either by ballistic flow into a feed tube, or by means of a capillary. A triangular target gas density distribution $\rho(z)$ is formed by the molecular flow of the atoms through the tube (*Source*: E. Steffens).

The straight beam tube has to provide sufficient space to allow the energetic stored beam particles to pass. In some cases, the beam during injection needs more space compared with that during the measurement which requires a cell to open, a delicate technical problem. Fortunately at the HERA electron ring, the injected beam was small enough that a cell of fixed size could be used, in contrast to a proton machine like the LHC. The feed tube is designed according to the dimensions of the free beam of the source. It should be noted that for an electron machine with its high level of X-rays, coating with Teflon is unstable and must be replaced by a different material.

For a complete polarized target, a source of polarized target atoms is required, injecting into the storage cell, as shown in Fig. 3.2. The Madison group used an Atomic Beam Source (ABS) injecting *in flight* the atoms, and this was taken up by the Heidelberg–Madison–Marburg–Munich group for the FILTEX target (see Section 3.6.2). Another method which has been investigated is to polarize hydrogen atoms by *spin-exchange* with laser-pumped potassium (K) vapor, the so-called Laser-Driven Source (LDS), proposed by Roy Holt from Argonne, later Illinois, and studied by several groups, incl. Argonne, Erlangen, Illinois and Duke Universities and BINP Novosibirsk, motivated by the very attractive estimates of potential target parameters. Whereas laser optical pumping of alkali vapor is well understood and very efficient, the use of a hot, coated storage cell with potassium vapor and the very reactive atomic hydrogen remained to be demonstrated. The parallel development of the ABS and the LDS targets continued during the early phase of HERMES running. A test setup of an LDS target built by Erlangen and Illinois was investigated in the Indiana Cooler Storage Ring, yielding unstable results, probably due to chemical reactions within the hot cell. This brought the LDS development to an end and the ABS target became the workhorse of the HERMES experiment in its successful data taking for a decade.

For the helium-3 target, the metastability exchange optical pumping (MEOP) technique, invented by King Walters, Laird Schearer and Forrest Colegrave at Rice University in 1963, was the obvious choice of the Caltech-MIT group (see Section 3.6.1).

3.2   North American Origin of HERMES

3.2.1   *Context*

The Kellogg Radiation Laboratory at Caltech in the mid-1980s was an intellectually exciting environment for a young physicist. Kellogg's traditional focus over many decades had been in the study of nuclear reactions that occur in stars and the lab was an international center for nuclear physics. The excited nuclear state in carbon predicted by Hoyle had been first observed in Kellogg. Frequent visitors included Hans Bethe, Gerry Brown, Hans Weidenmüller, and Ben Mottelson. In 1984, one of the leaders of Kellogg's nuclear astrophysics research, William (Willy) Fowler, had received the Nobel Prize in Physics. Willy's enthusiasm for life and physics was infectious. In addition, there were classes from Richard Feynman and Kip Thorne and a reading course from Roger Blandford. Ron Drever was starting up LIGO. In East Bridge, Felix Boehm's group carried out research into fundamental properties of neutrinos and future MIT physics faculty colleague and Physics Department Head, Peter Fisher, was a fellow Caltech graduate student there.

While this rich environment made for an inspiring education and training for a young physicist, the cutting edge research at Kellogg in the mid-1980s was focused on using the electroweak interaction to probe the nuclear force. The fundamental theory of the strong interaction, Quantum Chromodynamics described in Chapter 1, had been recently formulated and US nuclear physicists had decided that a new electron accelerator was to be built to study nuclei from the QCD perspective of quarks and gluons. In 1984, the Continuous Electron Beam Accelerator Facility (CEBAF), the new national user facility for electron scattering in the US, was launched at Newport News, VA. The design energy was 4 GeV, which was really too low to study quarks. In 1985, the design of the CEBAF machine was changed to a superconducting RF linac and construction began in 1987. The first physics experiment began production data taking in Hall C in November 1995. At Kellogg in the mid-1980s, under the leadership of Caltech faculty Steven Koonin, Robert McKecwn and Bradley Filippone, a new research direction, namely to understand nuclei in terms of quarks and gluons using the electroweak probe, was established.

Richard Milner defended his Caltech Ph.D. thesis on a search for free fractional charges using the Caltech Pelletron under the supervision of Robert McKeown in December 1984. Richard Feynman, Ronald Drever and Charles Barnes, together with Chairman McKeown, provided a 2 h grilling, with Feynman asking most of the questions. While he had post-doc offers from other groups, he decided to remain at Caltech and continue to work with McKeown on electron scattering experiments at SLAC and on development of a polarized ^{3}He gas target for electron scattering experiments. Originally, this was motivated by the possibility of experiments with internal gas targets in storage rings. The initial design of the CEBAF machine was based around a pulse stretcher ring and there was significant activity in the development of laser driven optically pumped polarized gas targets.

At the Kellogg lab at this time, there was a push to identify major new opportunities using electron scattering to study nuclei. Under Koonin's leadership, there were visits by the leading theorists in this field such as Dirk Walecka and William (Bill) Donnelly. McKeown and Filippone led the experimental effort and the Caltech group joined a new program, known as Nuclear Physics at SLAC (NPAS), at the Stanford Linear Accelerator Center which provided electron beams with energies up to 6 GeV for experiments with nuclei. Raymond Arnold and his group from American University formed the cornerstone of this program which attracted physicists from Argonne National Laboratory, University of Basel, University of Virginia, and other institutions. Milner worked as a Caltech post-doc on a number of experiments at SLAC and these formed his apprenticeship in learning electromagnetic nuclear physics. Peter Bosted, Steve Rock, Zen Szalata and others in Arnold's group were unique in being world experts on GeV electron scattering in the 1980s and it was the best place to learn the experimental and analysis techniques necessary to be successful at the future CEBAF facility. The students trained by Arnold's American University group are now leaders in the Jefferson Lab program. In addition to the 6 GeV program at the linear accelerator, there was another collaboration, involving many of the same physicists, called PEGASYS, considering possible experiments using internal gas targets at the PEP storage ring at SLAC. PEP could reach a maximum electron energy of 15 GeV, significantly higher than either NPAS or the planned CEBAF. Frank Dietrich and Karl Van Bibber from Livermore National Laboratory were leaders of the PEGASYS effort and the MIT nuclear theorist Arthur Kerman was a frequent presence at SLAC meetings. The principal scientific focus of PEGASYS was to understand the process by which a quark struck in a high-energy electron scattering event propagates through the nucleus and forms a detectable hadron. Motivated by the interest in experiments with internal gas targets, there were a number of workshops held at Stanford in the late 1980s.

Work began in 1984 at the Caltech Kellogg lab on the development of a polarized ^{3}He target for electron scattering experiments. Initially motivated by the desire to determine the charge elastic form factor of the neutron, Milner worked out that it could also be used to measure spin-dependent deep inelastic scattering from the neutron and reported this at the Princeton workshop in October 1984. In the years 1984–1987, the polarized ^{3}He gas target development continued. The principal focus was on the realization of a relatively thick polarized target that could be used with an external electron beam of energy of order 1 GeV, e.g. available at MIT-Bates, to determine the neutron elastic form factors.

In his years as a Caltech post-doc, Milner typically visited New England in the summer months to attend the nuclear physics Gordon Conference and to visit the CEBAF laboratory in Newport News, VA. The Gordon Conference was a perfect venue for a young physicist to become familiar with the research frontiers as well as the leading physicists. In the summers of 1984 and 1985, the CEBAF machine design consisted of a pulse stretcher ring and Milner was a member of the internal target working group, under the leadership of Roy Holt. When summer 1986 arrived, CEBAF had a new director, Herman Grunder, and a new machine design, namely the superconducting RF linac, and the pulse stretcher ring was no more. In hindsight, the bold decision by Grunder was the correct one for delivery of the high luminosity external beam. However, for experiments with polarized beams and targets, the internal target option offers significant scientific advantages over experiments using external beams.

In the mid-1980s, Kellogg Lab attracted some of the best young people from around the world. Douglas Beck, Wolfgang Lorenzon and Juerg Jourdan were experimental post-docs; Xiangdong Ji, Andreas Schäfer, Friedrich-Karl Thielemann and Karlheinz Langanke were theory post-docs; and Malcolm Butler, David Potterveld, and Martin Savage were fellow graduate students. The worldwide parity violation program to search for strange quark contributions to the proton's magnetism was initiated by Beck and McKeown in Kellogg at this time.

Living in Pasadena in the mid-1980s was stimulating beyond the physics. In large part driven by Willy Fowler, Kellogg Lab had an active social life that brought together faculty, staff and students. The annual Kellogg Halloween party was particularly memorable with imaginative costumes the order of the day. The speaker at the weekly Kellogg seminar at 4 PM on Friday faced an audience drinking beer. One particular distinguished speaker embarking on a description of his research attracted the comment from Willy: "I will need something stronger than beer to believe this!". Thom Moore and his band Train to Sligo played Irish folk music at the Loch Ness Monster Pub on weekends and this was a hangout for members of the Kellogg Lab. Moore was one of the great songwriters of the modern era in the Irish folk idiom. The Armenian-American artist Jirayr Zorthian and his wife Dabney, who lived on their ranch in Altadena, were regulars. Zorthian escaped the Armenian genocide as a child and studied art at Yale and in Italy in the 1930s. He was a friend of Richard Feynman, who also lived in Altadena. Zorthian and Feynman's attempts to teach other art and physics are described in the book *Surely You're Joking, Mr. Feynman!* The Pasadena Doo Dah Parade, the irreverent satire on the conventional Pasadena Rose Parade, was originated in 1978 by a group of friends sitting in the Loch Ness Pub. The parades transformed the staid suburban Pasadena streets, where a lone pedestrian was a rare sighting, into urban streets crowded with people for a few days each year.

In that summer of 1986, Milner was visiting an Irish friend who was painting houses in the Rockport, MA area. As the MIT-Bates lab in Middleton, MA was only about 20 min away, he called and arranged a visit. He met the Bates Director Ernest (Ernie) Moniz for the first time and the pleasant conversation about nuclear physics on a comfortable couch left a favorable impression. Moniz made it clear that MIT saw an opportunity in the decision by CEBAF to change its machine design and he was committed to realizing a program of internal target physics using polarized beams and targets at the Bates Lab. Moniz certainly provided encouragement to one young physicist at the right time to continue the development of polarized internal gas targets. Finally, Moniz welcomed a proposal to initiate measurement of spin-dependent electron scattering from polarized ^{3}He at his MIT-Bates lab using the existing external beam.

3.2.2 *1987: SLAC internal target workshop and first contacts with DESY*

In January 1987, a workshop on Electronuclear Physics with Internal Targets took place at SLAC. It was organized by a group representative of the scientific interests at SLAC, MIT, and Caltech. There were talks by leading figures interested in pursuing experiments with internal targets at storage rings. The opening talk was given by Terry Sloan from the University of Lancaster and the spokesman of the EMC experiment. His talk focused on meson production. Richard Milner gave a talk on the physics possibilities with a polarized ^{3}He internal target. In discussion between talks, Sloan made it clear that the EMC collaboration was finding a real surprise in the analysis of the spin-dependent muon-proton high energy scattering data. If there was a surprise on the proton, then the measurement on the neutron was strongly motivated as the proton and neutron were related by the Bjorken Sum Rule, a fundamental relation in the Standard Model. Sloan remarked that he thought that the idea of using polarized ^{3}He as an effective polarized neutron, which he had not heard before, was interesting. It was mentioned that the new HERA collider under construction at the DESY laboratory in Hamburg, Germany was being designed to have polarized electron and positron beams (Fig. 3.3).

Following the workshop, the PEGASYS group investigated the possibility to implement polarized electron beams in the PEP storage ring. With conservative assumptions about running parameters, a high quality measurement of the neutron structure function would have been straightforward to carry out. However, the SLAC management was negative. Milner decided to investigate the possibilities at HERA. He corresponded with Terry Sloan, who wrote in a letter dated 15 June 1987: "Concerning your enquiry about the use of HERA, it is clear that the higher energy is desirable to get down to

Fig. 3.3:   Aerial photograph of the DESY neighborhood of Hamburg, looking towards the Hamburg airport in about 1990. The overlaid white dashed lines indicate the accelerators HERA and PETRA. The East Hall, the location of the HERMES experiment, is indicated by the uppermost circle (*Source*: DESY).

the lower x but it is a tricky question (Fig. 3.4). I have briefly looked at doing the same thing at LEP. You have been misinformed about the HERA timetable. My colleagues here in the H1 collaboration inform me that there are no plans to put e^+ in HERA but to start in mid 1990 with $e^- - p$. However, they do plan to polarize the electrons. The main problem at both HERA and LEP is that the electrons are polarized transversely by synchrotron radiation and then a spin rotator is used to rotate to longitudinal and then back to transverse at each intersection. With the apparati planned for LEP and HERA there is no room for a gas jet target as they get the detectors in as closely as possible to the beam pipe. I think the large collaborations will be unwilling to move their gigantic apparati to accommodate a gas jet target so one has to find an appropriate slot to do the experiment when an apparatus is dismantled."

University of
LANCASTER
Department of Physics

Lancaster, United Kingdom LA1 4YB
Telephone: (0524)65201 Telex: 65111 Lancul G

15th June 1987

Dr. R. Milner,
California Institute of Technology,
452-48 Physics,
Pasadena,
California 91125,
United States of America.

Dear Richard,

Thank you for your letter and interesting plot. I enclose a preliminary draft
of our paper on the proton and copies of transparencies of a seminar to be
given at CERN by Vassili Papavassiliou — a Yale student who is writing a thesis
on our experiment. He computes the neutron asymmetry, assuming the Bjorken
sum rule in the transparencies.

Concerning your enquiry about the use of HERA, it is clear that the higher
energy is desirable to get down to lower x but it is a tricky question. I
have briefly looked at doing the same thing at LEP. You have been misinformed
about the HERA timetable. My colleagues here in the H1 collaboration inform
me that there are no plans to put e^+ in HERA but to start in mid 1990 with e^--p.
However, they do plan to polarise the electrons. The main problem with both
HERA and LEP is that the electrons are polarised transversely by synchrotron
radiation and then a spin rotator is used to rotate to longitudinal and then
back to transverse at each intersection. With the apparati planned for LEP
and HERA there is no room for a gas jet target as they get the detectors in
as closely as possible to the beam pipe. I think the large collaborations will
be unwilling to move their gigantic apparati to accommodate a gas jet target
so one has to find an appropriate slot to do the experiment when an apparatus
is dismantled.

I have a colleague at DESY who is working on the spin rotators for HERA. I
will discuss the idea with him. It will be better to do some ground work of
this sort before approaching the DESY management. I will also discuss the
idea with a colleague at CERN who is intimately involved with polarisation
studies to see if there is any feasible way of putting the target in LEP.

I go to the Uppsala conference next week and I will talk to people there. I
will write to you again when I return from Uppsala.

Yours sincerely,

Terry.

Dr. T. Sloan.

Fig. 3.4:   Letter on 15 June 1987 from Terry Sloan to Richard Milner.

HERA (German: Hadron-Elektron-Ring Anlage, English: Hadron-Electron Ring Accelerator) was a particle accelerator at DESY in Hamburg. It began operating in 1992. At HERA, electrons or positrons were collided with protons at a center of mass energy of 318 GeV. To date, HERA is the only lepton–proton collider in the world ever built. The principal scientific motivation for HERA was to search for particle sub-structure in the point-like quarks and leptons of the Standard Model. Also, it was on the energy frontier in certain regions of the kinematic range. HERA ceased operation on 30 June 2007. The HERA tunnel is located under the DESY site and the nearby Volkspark around 15–30 m underground and has a circumference of 6.3 km. Leptons and protons were stored in two independent storage rings on top of each other inside this tunnel. There were four interaction regions, two of which were used by the collider experiments H1 and ZEUS, which were conceived in the early 1980s to test the Standard Model. From the beginning, HERA was designed so that the stored positron or electron beam could became naturally transversely polarized through spin-dependent radiation (see Section 3.7). The characteristic build-up time expected for the HERA accelerator was approximately 40 min. Spin-rotator magnets were designed to rotate the beam spin into the longitudinal direction, a requirement for optimal measurements.

Sloan agreed to accompany Milner to Hamburg to introduce the idea to the DESY management. It was a long shot that a post-doc with a good idea could convince an international high-energy physics laboratory building a new frontier collider to seriously consider adding a fixed target experiment to study the spin of the proton and neutron. Thus, Milner bought his own Los Angeles–London plane ticket, courtesy of his sister Margaret who worked at that time for British Airways. Sloan and Milner met at Heathrow airport and traveled together to Hamburg. They met with DESY Director Volker Soergel and DESY Research Director Paul Söding in late September 1987 and discussed the idea to mount an experiment that would use the longitudinally polarized electron beam of the future HERA collider scattering from polarized internal gas targets of hydrogen, deuterium and ^{3}He. It was clear that both DESY representatives thought that the scientific question at issue was important and that an experiment should receive consideration. However, there were a number of critical issues identified: Where would the experiment take place? Who would form the collaboration? Who would provide the funding? How would you handle the high-energy proton beam in the experiment? How would you pump away the gas in the target? The East Hall was a likely candidate for the location and a visit was made there to inspect the many floors of empty rooms. The idea survived a discussion with Gustav-Adolf (Gus) Voss. The gas issue almost became a showstopper as the DESY vacuum group used helium to leak check the system. They objected to a helium gas target as it would eliminate a crucial diagnostic. However, after it was pointed out that the polarized target used ^{3}He and the vacuum group used ^{4}He and their diagnostic equipment could easily distinguish between the two isotopes, the issue evaporated. Milner returned to Pasadena with the understanding that a letter of intent to carry out the proposed experiment at HERA would be submitted for consideration by the DESY Physics Research Committee.

3.3   European Origin of HERMES

3.3.1   *Surprising results from EMC*

In autumn 1987, at a meeting of the EMC collaboration, the EMC results and the forthcoming publications were discussed together with possibilities for future spin experiments, especially with a polarized neutron target. At that meeting Geoffrey Court from Liverpool reported about the PEGASYS project at SLAC and also about rumors concerning the idea of US–American colleagues to perform such an experiment with internal polarized gas targets at HERA.

When Bogdan Povh learned about this, he was immediately electrified, since he realized that this was an ideal project for his division at the MPI-K. On the one hand, there was the group around Erhard Steffens, developing the internal polarized gas target for the FILTEX experiment, and on the other hand, there was the NMC group around Klaus Rith, with long-standing experience in DIS experiments.

Klaus Rith is an accomplished experimental physicist. He had analyzed part of the EMC iron data. He played a decisive role in the discovery of the nuclear EMC effect, by convincing his colleagues that the observed difference in the structure functions for iron and deuterium was not due to a systematic effect but originated in a new physical mechanism, namely a modification of the nucleons quark distributions due to the nuclear environment. He also drafted the first version of the corresponding publication. Klaus has been the first spokesperson of the NMC experiment, that studied the nuclear effects from a series of nuclear targets in great detail and also measured the proton and deuteron structure functions with unprecedented accuracy. In 1985, Bogdan Povh offered Klaus Rith a position as leading scientist at the MPI-K with the aim to develop together a better understanding of the nuclear force in terms of quarks and gluons. Rith played a central role in the HERMES experiment for the next 30 years. Bogdan Povh contacted Volker Soergel, chairman of the DESY board of directors, by phone. They concluded that it was not completely unrealistic to assume that such an experiment could become part of the HERA scientific program. Soergel encouraged Povh to let his group look into the details of such a project and eventually send a letter of intent to the DESY directorate and PRC.

After this rather positive answer, Bogdan Povh persuaded Klaus Rith to take over the leadership of the project. In early December, Klaus Rith visited DESY to discuss details with some key persons, in particular with Volker Soergel, Paul Söding, Friedhelm Brasse, Franz Eisele and R. Kose.

A week later, Klaus Rith gave a presentation at an NMC meeting at CERN, where he reported about this visit, posed some questions concerning the feasibility of the idea and also encouraged the colleagues from the various NMC groups to join the efforts of the MPI-K and sign the foreseen letter of intent.

Terry Sloan provided Richard Milner with copies of Klaus's transparencies. Richard immediately called Klaus and on 7 January 1988 he wrote a letter to him where he addressed and clarified some of the questions raised in Klaus presentation. At the end of the letter he wrote: "I would like to stress again that this experiment cannot come about without a strong European component to the collaboration and we would warmly welcome your group as collaborators. I look forward to hearing from you soon." After some lengthy considerations Bogdan and Klaus decided to not follow this invitation.

As a consequence, at the end of January/early February 1988 two separate letters of intent (LOI) were sent to the DESY directorate. The US LOI was signed by scientists from Caltech and Argonne and the European LOI was signed by groups involved in NMC and FILTEX (Amsterdam, Bielefeld, Delft, Freiburg, Heidelberg, Madison, München and Torino). These LOIs were discussed by the DESY PRC at its meeting on March 3rd and 4th. The PRC showed great interest in the idea but recommended that first a number of questions must be addressed: how can the technical challenges to installing such an experiment in one of the HERA halls be overcome and what are the possible interferences with the other HERA experiments? The PRC also requested that the two groups should join each other such that finally the PRC needed to consider only one single project. This information was formulated in a letter by Volker Soergel from 21 March, sent to both Richard Milner and Klaus Rith.

3.3.2   *Early years in Hamburg*

In 1967, Steffens was a student of the University of Hamburg and looking for a subject for his Diploma thesis. For the previous 3 years, he was, in parallel to his studies, a member of different rowing teams of Hamburg University coached by Walter Schröder, Gold medal winner of the German Eight at the Olympic Games in Rome 1960. In 1967, Steffens and his team were second in the German Championship of the light-weight Eight competition, and he decided to focus on a career in science.

Physics in Hamburg at that time took place at the downtown campus with applied, theoretical and experimental physics, and the growing research center DESY for electron scattering at the city boundary, founded in 1959. Among the faculty members were Pasqual Jordan and Harry Lehmann (Theory), and experimentalists Willibald Jentschke, founder of DESY, Gustav-Adolf Voss, and Björn Wiik until his tragic death in 1999. After looking around, Steffens joined a group working on the development of a spin-polarized lithium ion beam — in retrospect a far-reaching decision! The group was located in the First Institute for Experimental Physics (IEP-1) chaired by a single professor, Hugo Neuert. In IEP-1, there were many small groups active in atomic and nuclear physics and heavily involved with teaching, lab courses, etc., which provided a modest income for the research students. Such an institute with a single Ordinary, a Chair professor and director, had the typical structure of a university institute at that time before the students revolution in 1968, which forced the introduction of the US departmental system. This was pioneered in Germany by the *Physik Department* at the Technical University Munich (TUM) in 1965.

The principal experimental technique employed by the IEP-1 group was the Molecular Beam Method invented by the French physicist Louis Dunoyer and perfected by Otto Stern during his time (1923–1933) as founding director of the new Hamburg Institute for Physico-Chemistry in the Jungiusstrasse, next to the IEP-1.[8] Unfortunately, being Jewish, Stern was forced to emigrate to the USA in 1933. His most famous paper was with Walter Gerlach on the quantization of the spatial orientation of angular momenta, a work performed in Frankfurt in 1921–1922, and which is described in Section 2.2. Stern was awarded the 1943 Nobel Prize in Physics for his development of the molecular beam method and the first measurement of the magnetic moment of the proton. Stern followed a strategy laid out in 1926 to develop the Molecular Beam method systematically, documented in a series of # Untersuchungen zur Molekularstrahlmethode (U.z.M.) papers in Zeitschrift für Physik. In 1933 in U.z.M. 24, he described with Frisch the first results on the magnetic moment of the proton by deflecting a cold beam of Ortho-Hydrogen molecules, i.e. molecules with zero electron spin and two proton spins parallel — probably the most admirable result of Stern's work. The result for the g-factor of the proton showed a large deviation from the Dirac-value $g = 2$ for a point-like particle of s $= 1/2$, in retrospect the first indication of an internal structure of the proton.

In the 1960s, when Steffens joined the IEP-1, experimental methods developed in Stern's institute included velocity selectors, Langmuir-Taylor detectors for alkali beams or strong inhomogeneous fields for deflecting paramagnetic atoms like lithium-6 were still routinely used in the IEP-1 groups and workshops. The beams were generated by evaporation from an oven at about 1000°C. Isotopically pure lithium-6 metal, used for producing the fuel of the hydrogen bomb, could be obtained for research purposes from Oak Ridge National Laboratory, Tennessee, USA.

Led by Hilmar Ebinghaus and with Uwe Holm, a prototype polarized source of negative lithium-6 ions was built to allow injection into the electrostatic tandem accelerator of the Heidelberg Max-Planck-Institute for Nuclear Physics (MPI-K). From his Industry Practical in 1963 at ZEISS (Oberkochen, Würtemberg), Steffens had acquired a good knowledge of precision mechanics which he put to use throughout his career in all technical aspects of his projects. He became a member of the Hamburg–Heidelberg collaboration, with Dieter Fick, head of the MPI-K Heavy Ion group. The program of polarized heavy ions was conducted for more than a decade, later as a Heidelberg–Marburg collaboration, when Fick had accepted a Marburg chair position. Beams of vector and tensor polarized lithium-6, lithium-7 and sodium-23 beams could be accelerated by the EN and MP tandems and the RF booster. These unique beams were used to study shape effects in the fusion of aligned heavy ions in the vicinity of the Coulomb barrier. Lithium-7 and sodium-23 beams have nuclear spin-$\frac{3}{2}$ enabling for the first time scattering experiments with third rank spin polarization. In the 1980s, the ion source was further improved to include optical pumping of alkali atomic beams, enabling population and detection of the eight individual substates of the lithium-7 and sodium-23 beams.

• **The Antiproton Storage Ring LEAR at CERN and the FILTEX proposal:** In 1984, Steffens moved with his family to the small village of Challex near CERN, on the French side of the border, into a house of a CERN accelerator physicist who was on leave for a year at SLAC. There, he spent a full sabbatical year at CERN in the LEAR machine group to gain insight into the accelerator technology of storage rings. LEAR, the Low Energy Antiproton Ring at CERN, was an innovative facility to store and cool *antiprotons*, the anti-particle of the proton with negative charge. Beams of antiprotons could then be supplied to different experiments. An internal target experiment was also being prepared. One of the LEAR experiments was PS 173, elastic scattering of antiprotons from a liquid hydrogen target, strongly supported by the MPI-K division of Bogdan Povh. In discussions with Povh, Thomas Walcher and LEAR machine physicists, in particular Dieter Möhl, expert on beam dynamics and phase space cooling and mentor of Steffens during his time at CERN, a proposal for a new internal-target experiment was developed: the **FIL**ter **T**arget **EX**periment (FILTEX).

Alan Krisch (Michigan), then chair of the ISPC, called for one of his famous topical workshops, on Polarized Antiprotons (1985) at Bodega Bay, CA, in a motel which in 1963 provided the scenery for Hitchcock's film "The Birds". Among the participants of the workshop were Nobel Prize winners Owen Chamberlain (Berkeley) and Simon van der Meer (CERN), and other distinguished colleagues, like C.D. Jeffries, J.D. Jackson, D. Kleppner, and T. Niinikoski. In the discussion of the about 15 ideas, the FILTEX method was rated as "most practical" for creating stored polarized anti-protons (Fig. 3.5).

The goal of the FILTEX proposal was to polarize anti-protons stored in LEAR by spin-dependent attenuation on a polarized hydrogen target and to measure their (double-polarized) scattering on the same target. A central part of the proposal was to develop a polarized proton gas target of high polarization and two orders of magnitude higher density than available at that time. With Steffens as Spokesperson, the FILTEX experiment was proposed by a Heidelberg–Marburg–München–Houston–Madison–Rutgers collab-oration to the PSSC committee of CERN in 1985 and accepted conditionally, subject to a demonstration of the new technique of

Fig. 3.5: Photo from the Workshop on Polarized Antiprotons at Bodega Bay, CA in 1985; organized by Alan Krisch (center). Right: Ken Imai (Kyoto); left: Erhard Steffens (Heidelberg) (*Source*: A.D. Krisch).

Spin-Filtering with protons in the Test Storage Ring (TSR), which was under construction at MPI-K.

This institute, located half-way up to the Königsstuhl, a mountain overlooking Heidelberg, with its Nuclear and Astrophysics departments, goes back to the tradition of Walter Bothe (Nobel Prize in Physics in 1954 for the coincidence method) of the Heidelberg MPI for Medical Research and the first German Cyclotron. It had two Tandem Accelerators, including the MP Tandem equipped with an RF post-accelerator. The laboratory workshops were able to handle large projects. A significant part of the equipment of the lab was designed and constructed in-house.

The next 5 years after the FILTEX proposal were filled with activities on design and construction of the target, largely driven by Gerhard Graw and Peter Schiemenz at LMU/TU Munich, using the resources at the Physik Department (at Garching near Munich), and Steffens' own group at MPI-K, with Hans-Günter Gaul (later Thyssen-Krupp), Friedemann Stock (later Stihl company) and

Kirsten Zapfe, later at DESY. The group of Dieter Fick of the University of Marburg, a long-standing collaborator, also worked on the target project. This included Frank Rathmann, who wrote a thesis on the experimental demonstration of Spin-Filtering (see Section 3.6.2), now at FZ Juelich, and Wolfgang Korsch, now at U. Kentucky, who wrote a thesis on the velocity distribution of the atomic beam.

A huge effort of the MPI-K technical staff went into the design and construction of the new storage ring TSR. This was led by Eberhard Jaeschke, with strong contributions from Dieter Habs, Dieter Krämer and Roland Repnow. Later Jaeschke initiated the new synchrotron radiation source BESSY II in Berlin-Adlershof, the new research campus of the Humboldt University. The TSR was optimized for the storage of heavy ion beams in different charge states, and for laser cooling experiments. Steffens took responsibility for the beam position measurement system and for the DC-transformer which provided a non-destructive measurement of the beam current. The technology followed CERN standards which was developed for the anti-proton source, including vacuum-fired, bakeable chambers for very low pressure. As the FILTEX target was designed to operate within the LEAR storage ring, these standards were also applied to the target.

In 1991, the workshop on "Polarized Gas Targets for Storage Rings" took place at the MPI-K in Heidelberg with 61 participants from 12 countries, representing the majority of experts in this field. Figure 3.6 shows the participants. First results on the parameters of the FILTEX target were reported.

3.3.3   *Particle physics in Europe in the 1980s*

Research at CERN in the early 1980s was directed towards the production of the Intermediate Vector Bosons Z^0 and $W^{\pm}$, see Fig. 1.4. Carlo Rubbia and Simon van der Meer were awarded the 1984 Nobel Prize in Physics for their discoveries by colliding protons and anti-protons, counter-rotating in a single 450 GeV storage ring, the Super Proton Synchrotron (SPS) collider. Positively charged

Fig. 3.6: Participants of the 1991 workshop on Polarized Gas Targets for Storage Rings, including W. Haeberli, T. Clegg, D.G. Crabb, L. W. Anderson, T Gentile, W.A. Kaufmann, E. Kinney, R. Pollock, R. Raymond, T. Wise (USA); A. Belov, V.E. Kuzik, D.M. Nikolenko, Y. Pilipenko, D. Toporkov (Soviet Union); G. Arduini, G. Cinque, L. Dick, A. Penzo (Italy); J.L. Lemaire, A. Zghiche (France); G. Court (UK); K. de Jager, H. de Vries, J. van den Brand (Netherlands); Y. Mori, Y. Wakuta (Japan); W. Gruebler, W. Kubischta, G. Pavia, P. Schmelzbach, B. Vuaridel (Switzerland); D.P. Barber, G. Clausnitzer, M. Düren, D. Fick, G. Graw, W. Heil, M. Mertig, K. Rith, P. Schiemenz, K. Zapfe (Germany) (*Source*: MPI-K Heidelberg).

protons and anti-protons, with their opposite negative charge, can circulate within the same channel of magnetic lenses in opposite directions. Therefore, the existing SPS could be employed as a proton–anti-proton collider at moderate extra costs. The price to pay was to develop an intense source of anti-protons which did not exist. Its development was a milestone in accelerator technology and required the development of phase space cooling, i.e. a process to reduce the "beam size" of particles in a storage ring by the action of RF correction signals.

The anti-proton program was terminated in favor of constructing and operating the Large Electron and Positron collider (LEP), built in a tunnel of 27 km in circumference and designed as a Z^0 and $W^\pm$ factory for a precision study of their properties. In parallel, the 450 GeV SPS was employed for the production of pions, muons and kaons. High-energy muons, which can be regarded as heavy electrons, are produced by in-flight-decay of energetic pions. Due to their origin in a decay mediated by the Weak Force (see Section 1.1.3.3), they are polarized along the beam direction (longitudinal polarization). Muon beams were used by the EMC collaboration to study structure functions reflecting the sub-structure of nucleons made up of quarks and gluons. When targets of different nuclei were used, a mass dependence was observed, called *the EMC effect*. Muon beams at CERN are discussed in Section 6.4.

A new observable became accessible by employing cryogenic targets at very low temperature and high magnetic field with a certain fraction of polarized protons. In combination with polarized muon beams, a new spin-dependent structure function could be extracted. These new measurements of the EMC experiment were interpreted in terms of the quark polarizations in a polarized proton, i.e. how much of the proton polarization is carried by quarks. To their surprise the researchers found a rather low fraction in contrast to current models, triggering the so-called *spin crisis* and a number of proposals to go beyond these results (see Sections 2.4 and 2.5).

• **Two extreme options for double-polarized DIS measurements:** As explained in Section 1.4, most of the information on the structure of sub-atomic particles is derived from scattering experiments $a \rightarrow b$ where a beam of particles a (e.g. electrons, protons, α-particles) of intensity I (particles/s) hits a target b of areal density t (target atoms/unit area). Then, the scattering rate is proportional to the luminosity $\mathcal{L} = I \cdot t$, depending on beam and target properties. The rate R_X for a specific process $a + b \rightarrow X$ is given by $R_X = \mathcal{L} \cdot \sigma_X$, with σ_X being the cross section of process X. A fundamental quantity in discussing the scattering of sub-atomic particles is the unit of cross-section, the *barn* $= 10^{-24}\,\mathrm{cm}^2$ (see Appendix A.2). This unit was invented in neutron physics where a barn is regarded as huge ("as big as a barn", see Appendix A for a discussion). It is widely used in particle physics with its much smaller cross-sections (milli-barn, micro-barn, nanobarn, etc.)

A certain luminosity $\mathcal{L} = I \cdot t$ can be realized in different ways between the two limiting cases:

(1) A low intensity polarized beam, e.g. muons from decay of pions, on a thick solid polarized proton target, or
(2) An intense beam, e.g. electrons stored in a ring, impinging on a thin gas target.

The EMC experiment at CERN and its later versions SMC and COMPASS represent the first option. A beam of spin-polarized muons of about $100\,\mathrm{GeV}$ in energy hits a 1-m long target at temperatures below $1\,\mathrm{K}$ containing polarized protons. The HERMES experiment, proposed in 1989 in the wake of the spin crisis, represents the second option. The successful concept of the HERMES experiment was based on two innovative new developments:

- *Self-polarized electrons of high-energy*: Electrons very close to the velocity of light circulating in a storage ring show a slow build-up of vertical polarization via the Sokolov-Ternov effect; see Section 3.7.
- *Polarized gas targets*: Polarized atomic gas targets of high purity and polarization, consisting of hydrogen, deuterium, or helium-3, a rare light isotope of helium. The polarized helium-3 target used by HERMES was based on optical pumping and was developed by a MIT-Caltech collaboration led by Milner, and is introduced in Section 3.6.1. The polarized hydrogen/deuterium target was developed at the MPI Heidelberg in preparation of the FILTEX experiment by a Heidelberg–Madison–Marburg–Munich–Madison–Rutgers collaboration, led by Steffens, and is described in Section 3.6.2.

3.4 1988: HERMES is Born

In October 1987, work began on developing a letter of intent from an Argonne–Caltech collaboration. It was proposed to measure the deep-inelastic spin-dependent structure functions for the proton and neutron using internal polarized targets and the longitudinally polarized 30 GeV electron beam in the HERA electron storage ring. The polarized internal gas target would consist of laser driven optically pumped sources of hydrogen, deuterium and ^{3}He feeding a storage cell through which the electron beam passed. An important issue was the shielding of the proton beam from the magnetic field of the spectrometer magnet. It was proposed to employ a superconducting tube, which had been successfully implemented in an experiment led by Louis Osborne at SLAC in the early 1970s. In writing the letter of intent, it became clear that the experiment was an unique opportunity to technically realize internal polarized internal gas targets of polarization and thickness motivated by a fundamental physics problem.

In March 1988, Richard Milner attended the Lake Louise Winter Institute in Canada and met Bogdan Povh, Director at the Max Planck Institute für Kernphysik (MPI-K) in Heidelberg, Germany. Povh informed Milner of the second letter of intent to carry out measurement of spin-dependent high-energy electron scattering from polarized internal gas targets at HERA, led by his Heidelberg colleague Klaus Rith. Later that month, Milner interviewed for a faculty position at the University of Wisconsin-Madison and met Willy Haeberli, one of the pioneers of the development of the storage cell and atomic beam source technology. Willy was one of the collaborators on the second letter of intent. Subsequently, Milner received an offer of a junior faculty position at Madison. Further, he had interviewed for a similar faculty position at MIT, but the position was offered to another candidate. Milner was all set to join Haeberli at Wisconsin when MIT came back with an offer: their first choice candidate had declined. After some consideration, Milner chose to take up a faculty position at MIT in July 1988.

In May 1988, at the Third Intersections Conference in Rockport, Maine many of the principals behind the two letters of intent to DESY attended and there was a meeting to discuss how to proceed. This is the location and date where a collaboration to mount a new experiment to study the spin structure of the nucleon using polarized internal gas targets at the HERA storage ring was born. It was agreed to work together as a transatlantic collaboration with the DESY management to propose this new experiment. At the Rockport, Maine meeting Vernon Hughes from Yale, one of the originators of using spin-dependent DIS to study the origin of the proton spin, was in attendance. In discussion in the plenary session, Hughes made it clear that he was skeptical of the technical feasibility of the HERA experiment.

The meeting was very informal. From the US side most of the proponents were present, including Costas Papanicolas from Urbana. Most importantly, also Otto Häusser, from the TRIUMF laboratory in Canada, participated in the discussion and expressed interest on behalf of several Canadian institutes to join the collaboration. This was the start of the Canadian involvement. Klaus Rith and Willy Haeberli were the representative of the European LOI. Furthermore Bernard Frois from CEN Saclay took part in the discussion. The participants agreed to follow the request of the DESY PRC and to form one collaboration, with Richard and Klaus as co-spokespersons.

In August 1988, a document *Feasibility Studies for an experiment to Measure the Spin Dependent Nucleon Structure Functions at HERA* was prepared by a collaboration of physicists involving the proponents of the two letters of intent from Argonne, Caltech, MPI Heidelberg, U. Illinois at Urbana-Champagne, U. Wisconsin-Madison, U. Marburg, MIT, U. Munich, U. Laval, Saclay, U. Torino, TRIUMF and U. Uppsala. Quite a number of these studies were done by Michael Düren, who had started early in 1988 as postdoc at MPI-K. The "Feasibility studies" were presented to the DESY PRC by Klaus Rith on 9 September 1988. Official letters dated 18 October 1988 from DESY Research Director Paul Söding were sent to both Richard Milner and Klaus Rith. The letter to Milner is shown in Fig. 3.7, where he reported on the consideration of the document by the DESY Physics Research Committee.

The committee agreed that there was great physics interest in such an experiment. Also, the present collaboration was felt to to be qualified to develop and realize the techniques of the polarized targets and the detector. DESY had looked into the possibilities for accommodating the experiment at HERA. The fact that the proton

DEUTSCHES ELEKTRONEN — SYNCHROTRON DESY

NOTKESTR. 85 · 2000 HAMBURG 52 · TEL. 040/89 98-0 · TELEFAX 89 98—32 82 · TELEX 2 15 124 desy d

Dr. R.G. Milner
M I T
Cambridge, MA 02139
USA

Hamburg, October 18, 1988

Dear Dr. Milner :

With this letter, I would like to summarize DESY's present position concerning the measurement of spin-dependent structure functions at HERA, following the recent discussions in the Physics Research Committee. This discussion was based on the Feasibility Study presented by you and your colleagues.

The committee agreed that there is great physics interest in such an experiment. Also, the present collaboration was felt to be qualified to develop and realize the techniques of the polarized targets and the detector. DESY has looked into the possibilities for accommodating the experiment at HERA. The fact that the proton beam has to be spatially well separated from the electron beam in the region of the experiment will make a major rearrangement of an interaction region necessary. This requires further study.

A basic question which must be answered before the experiment can be approved, is whether a sufficient degree of longitudinal beam polarization will be obtained at HERA. It is understood that the degree of longitudinal polarization should be at least 50% in order to make the experiment worthwhile. DESY will make a strong effort to achieve this, however definite results cannot be expected before spring 1991 at the earliest, after installation and commissioning of superconducting cavities and the spin rotator. Both items are being constructed at present.

Therefore, the collaboration is encouraged to submit a full proposal for evaluation by the Physics Research Committee. A decision for approval will have to be pending until the question of electron polarization in HERA is experimentally clarified.

Sincerely Yours,

P. Coding
(P. Söding)

Fig. 3.7: Letter on 18 October 1988 from DESY Research Director Paul Söding to Richard Milner.

beam has to be spatially well separated from the electron beam in the region of the experiment required a major rearrangement of an interaction region.

A basic question which had to be answered before the experiment could be approved, is whether a sufficient degree of longitudinal beam polarization would be obtained at HERA. It is understood that the degree of longitudinal polarization should be at least 50% in order to make the experiment worthwhile. Söding committed DESY to make a strong effort to achieve this.

In conclusion, the collaboration was encouraged to submit a full proposal for evaluation by the Physics Research Committee. A decision for approval would have to be pending until the question of electron polarization in HERA was experimentally clarified. At the Santa Fe meeting in October 1988 the Los Alamos group, represented by Helmut Baer, expressed interest in joining the collaboration. Indeed, the members of this LANL group became co-authors of the HERMES proposal, their envisaged contribution being the magnet and the electromagnetic calorimeter. Unfortunately both Helmut Baer and N. Tanaka died of cancer and LANL left the collaboration again in early 1992.

After his plenary talk at the Bonn High Energy Spin Physics Conference in September 1990, Klaus Rith was approached by Robert Avakian and Hamlet Vartapetyan from the Yerevan Physics Institute and Stanislav Belostotsky from the Leningrad Nuclear Physics Institute. They expressed their interest in joining the collaboration. After the LANL group had dropped out, some of its responsibilities were passed on to these two institutes.

At the end of the 1980s, there was significant activity in Europes hadron physics community to formulate the case for a future accelerator facility. Klaus was heavily involved in these activities and in close touch with the group leaders of many institutions and, in particular, Peter de Witt Huberts from NIKHEF, Enzo De Sanctis from Frascati and Salvatore Frullani from Rome. These finally joined the collaboration with their groups early in 1993.

In summer 1988, Milner and his wife moved from Pasadena, CA to Boston, MA to take up a junior faculty position in the

Physics Department at MIT. The MIT position had been made available in conjunction with the approval and funding of the South Hall Ring at Bates. With Moniz as Director, MIT-Bates aimed to mount a program of spin-dependent electron scattering from polarized internal few-body gas targets at medium energies to pursue precision measurement of observables such as the electric form factor of the neutron. Two experiments had been approved to make the first measurements of spin-dependent electron scattering from polarized ^{3}He at Bates using gas targets employing optical pumping. One experiment employed the target at Caltech. Upon arrival at MIT, Milner immediately initiated a project to develop a polarized ^{3}He internal gas target for both the HERA and Bates South Hall Ring. The effort was located in a warehouse at the Bates Laboratory and MIT graduate student Kevin Lee joined in September. Upon the advice of Moniz, a newly graduated NIKHEF Ph.D. Johannes (Jo) van den Brand was hired by Milner and arrived with his family in Boston at the end of 1988.

3.5   Conceptual Description of the **HERMES** Experiment

To measure spin-dependent high-energy electron (or positron) scattering from the proton, both the electron and proton must be polarized with the electron spin being directed longitudinally, i.e. along the beam direction, and the proton spin being either longitudinal or transverse. The scattered electron is detected in a magnetic spectrometer and the scattering asymmetry, i.e. the difference in scattering rates divided by the sum, determined as a function of time. For an effective experiment, the beam and target polarization should be at least 50% and running times of order 1–2 months are essential to determine asymmetries of order 1% with adequate precision. Prior to HERMES, all experiments had involved targets where the polarized proton is embedded in a large amount of extraneous material. This has the consequence that in experiments where only the scattered electron is detected that the asymmetry is significantly diluted. Further, if hadrons are additionally detected in coincidence,

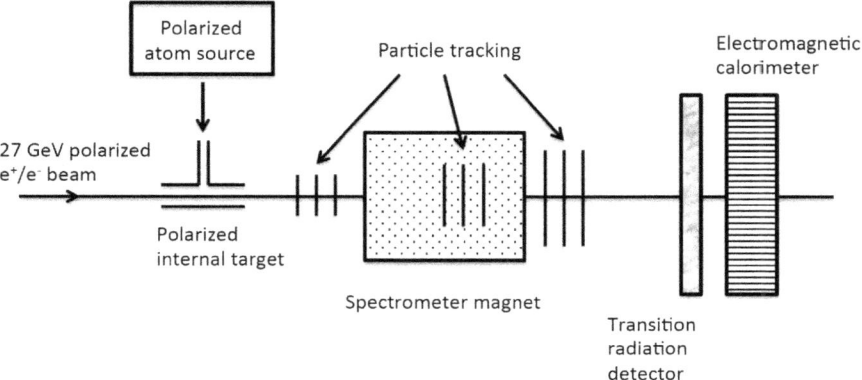

Fig. 3.8:   Schematic layout of the HERMES experiment (*Source*: R. Milner).

the dilution becomes significantly dependent on the kinematics of the experiment. Thus, HERMES offered the very attractive possibility to realize for the first time the desirable idea of pure lepton-nucleon scattering made possible by the polarized internal gas target.

Figure 3.8 is a schematic layout of the major components of the HERMES experiment. The high energy (27 GeV) longitudinally polarized electron e^- (or positron e^+) beam circulates in the 6.3-km circumference HERA tunnel beneath the suburbs of Hamburg. The beam passes through a thin-walled tube about 40 cm in length, called the *storage cell*, in which a significant pressure of polarized atoms builds up. The polarized atoms are fed from a source: an atomic beam source in the case of hydrogen and deuterium and a laser-driven source in the case of ^{3}He. The interaction of the polarized beam with the polarized atoms results in scattered beam particles and associated hadrons that are detected in the HERMES spectrometer. The spectrometer includes a magnetic field which both rejects uninteresting background and also affects the trajectories of the charged particles of interest that result from the beam–target interaction. The trajectories of the charged particles can be reconstructed in the particle trackers and thus their momentum can be deduced. Towards the rear is the transition radiation detector, which enables good particle identification and the electromagnetic calorimeter which measures the total energy of the scattered electron or positron. An electronics circuit triggers the detectors when interesting events are recorded and the complete information on the event is digitized and stored in a memory disk. A fraction of the recorded data is analyzed in real time as the experiment takes data to ensure that the system is functioning as designed and that the scattering asymmetry being determined is reasonable.

There are essential elements of the experiment missing from Fig. 3.8 to ensure clarity. There is a large differential vacuum pumping system on either side of the internal gas target to ensure that the target gas does not enter the high vacuum of the storage ring. In addition, there are both longitudinal and transverse beam polarimeters near the interaction region to precisely determine the polarization of the beam. Finally, there are diagnostic elements sampling the properties of the polarized atoms in the storage cell to ensure optimal performance.

3.6   Development of Polarized Internal Gas Targets

The use of gas targets internal to a storage ring for sub-atomic physics scattering experiments was pursued at several accelerator laboratories, including Frascati, Italy and Novosibirsk, Russia. The clear advantage is that the particles circulating in the ring are reused many times and may become polarized during storage. The race to higher energies has led to the development of powerful colliders that are known for their discovery of new particles, e.g. the antiproton-proton collider and LEP for the production and study of the Z^0 and $W^\pm$, the Large Hadron Collider (LHC) for the Higgs, and the HERA electron-proton collider for the study of the gluons that mediate the strong force. Circulating beams in the storage rings of colliders can be used as the projectile beam for a scattering experiment from an internal gas target at a lower effective energy. This opens up new scientific opportunities beyond those originally conceived for the collider. With careful design, such an internal target program can be carried out in parallel with the collider program. The internal gas target must be thin to avoid serious degradation of the stored beam for collider operation. Such programs were performed in the 1980s at the SPS collider at CERN (UA6 experiment) and the Tevatron, and presently at RHIC and at the LHCb experiment employing one of the 7-TeV beams of the LHC (see Section 7.4). Here we focus on the HERA electron ring with nominal energy 30 GeV.

At DESY in the 1980s, the 6.3-km circumference HERA collider was under construction, consisting of a 30-GeV electron ring, a 920-GeV proton ring and the two collider experiments H1 and ZEUS. About 25% of construction funds were provided by international partners (through in-kind contributions, including manpower). Ten countries from Asia, North America and Europe contributed components such as RF systems, magnet measurements and controls, superconducting dipoles, quadrupoles and correction magnets, and beam dumps. Two countries (China and Poland) contributed mainly through manpower performing work on various machine components. This successful way to build an accelerator by international partners became known as the "HERA-Model".

3.6.1 *Polarized neutron (^{3}He) target*

As described above, the development of polarized ^{3}He gas targets motivated by their utility as an effective polarized neutron target for scattering experiments got underway in 1984. CEBAF had been approved as a pulse stretcher ring and there was an internal target working group led by Roy Holt. A workshop on polarized targets took place at Argonne in May 1984. Bob McKeown attended and returned with the conviction to initiate a development effort at Caltech. By the end of 1984, Milner had completed his Ph.D. thesis and was working on the project. Graduate student Cathleen Woodward (Jones) was also a member of the team. The Caltech group made the choice to use the technique of metastabilty exchange optical pumping, developed in 1963 by Colegrove, Schearer and Walters. Targets and ion sources using flash lamps had been developed and used at low energy nuclear physics laboratories, including Basel, Caltech and Rice. However, by 1984 there was the promise of powerful, tunable new laser systems with the goal of attaining higher polarizations and polarization rates.

Michèle Leduc at École Normale Superiéure in Paris and Laird Schearer at the University of Missouri-Rolla and their colleagues were leaders in the development of these new laser systems. An alternative polarization scheme, known as spin exchange optical pumping, was pursued under the leadership of Timothy Chupp, then at Harvard University. This technique had the advantage that it could produce significantly higher density polarized ^{3}He gas.

An important meeting that brought together all active in polarized ^{3}He research took place at Princeton University in October 1984. Figure 3.9 is a photograph of the attendees. Richard Milner presented an initial estimate of what we could be learned about the quark contribution to the neutron spin structure in high-energy electron scattering from polarized ^{3}He.

Fig. 3.9: Photo of the attendees at the Princeton workshop in October 1984 on polarized ^{3}He sources and targets. Front (left to right): E.P. Wigner, R.J. Slobodrian, F.P. Calaprice, S.D. Baker, T.B. Clegg, R.D. McKeown, M.S. Dewey, T. Chupp. Middle: M.M. Lowry, R. Sherr, J. Powelson, J.M. Daniels, J. Dupont-Roc, D.P. May, R. Roy, J.G. Alessi, J.D. Brown, A.B. McDonald, S. Oh. Back: M. Schneider, R.T. Kouzes, R.W. Dunford, R.G. Milner, C. Rioux, W.H. Moore, R. Knize, W. Happer, J. Giroux, F. Laloe, D.E. Murnick (reproduced from the Proceedings of the Workshop on Polarized ^{3}He Sources and Targets at Princeton University, October 1984, with the permission of AIP Publishing).

At Caltech, it was decided to focus on development of a polarized ^{3}He gas target for a planned measurement of spin-dependent electron scattering from polarized ^{3}He at energies of 500–800 MeV. If successful, it was planned to propose an experiment at the MIT-Bates user facility which could deliver the required polarized electron beam. The first phase involved construction of a discharge lamp which produced 2% polarization in a cell made of copper and glass through which a charged particle beam could be passed. This enabled an important study at the Caltech Pelletron with low energy protons of beam depolarization of the polarized gas. Further, a model of the depolarization mechanism was developed which highlighted the important role played by the diatomic, singly-charged ion. Even though the polarization was low, the work encouraged the Caltech group to proceed with the development of laser systems. In the summer of 1985, Milner and his advisor McKeown spent several weeks at the Virginia Associated Research Campus in Newport News, VA and took part in the CEBAF Summer Study Group working on the scientific program for the future CEBAF. They reported on the Caltech development of a polarized ^{3}He target.

In January 1988, Milner successfully defended a Caltech proposal to carry out a measurement of spin-dependent electron scattering from polarized ^{3}He at Bates. This gave additional impetus to the development of a double-cell system where the ^{3}He gas was polarized at room temperature in a glass cell which was in diffusive contact with a copper cell kept at low temperatures to increase the density.

As described earlier, Jo van den Brand, Kevin Lee and Richard Milner began work on development of a polarized ^{3}He internal gas target for both the Bates South Hall Ring and the HERMES experiment in early 1989. The work was carried out at the MIT-Bates center in Middleton, MA. A flowing system of ^{3}He gas with 50% polarization at flow-rates in excess of 10^{17} atoms/s was established within about a year. With neither the Bates or HERA rings likely to become accessible in the near future, the group looked for other scientific opportunities to utilize the recently-developed target.

Such an opportunity arose in 1990 at the Indiana University Cyclotron Facility (IUCF) Cooler ring. The IUCF physicists had implemented a low-energy version of Simon van der Meer's Nobel Prize winning technique using electron beams. In addition, IUCF had a polarized proton source so that they could store intense beams of polarized protons at energies of several hundred MeV. Thus, was born Cooler Experiment 25 (CE-25): a measurement of quasielastic spin-dependent scattering of polarized protons from polarized helium-3. IUCF physicist James (Jim) Sowinski was a leader of this effort along with van den Brand, now a faculty member at the University of Madison-Wisconsin, and Milner. The scientific goal was to demonstrate that the polarized helium-3 nucleus was an effective polarized neutron target for a scattering experiment. Previous measurements at TRIUMF by Otto Hausser and his colleagues were inconclusive.

The target used a glass pumping cell engineered by MIT engineer Jim Kelsey to adapt to a custom-built aluminum vacuum chamber. The system had all-metal seals, was bakeable and the target polarization was rapidly reversible. The storage cell was made of aluminum and operated at room temperature. The target performed stably and reliably at a high level of performance throughout the data taking in 1990 and 1991.

CE-25 was the first internal gas target experiment where both the beam and target were polarized. The spin-dependence of both proton and neutron knockout reactions was measured. Using kinematic cuts to validate the knockout mechanism, it was definitively shown that the polarized helium-3 nucleus was a very good polarized neutron target. Further, the proton knockout measurements directly demonstrated a small ($\sim 1\%$) component of the ground state wave function for the first time. Kevin Lee from MIT and Michael Miller from Madison wrote excellent Ph.D. theses on CE-25.

With the conditional approval of HERMES in 1992, work at MIT focused on development of a polarized helium-3 target for the HERA electron storage ring. The successful target operation in CE-25 at IUCF meant that the polarized helium-3 source was relatively straightforward. However, the optimal storage cell for HERMES required significant development. Because of the lower cross section at HERMES, the target cell had to be cooled to low temperatures and the transverse dimensions had to be minimized, consistent with the beam halo characteristics.

3.6.2 *Polarized hydrogen/deuterium target*

In 1985, the FILTEX collaboration (Heidelberg–Madison–Marburg–Munich–Rutgers) started to work on the realization of the proposal approved by CERN conditionally, subject to the successful development of a polarized gas target (PGT) with suitable density and polarization. The goal was to demonstrate spin filtering with protons. As explained in Section 3.1, a storage cell target consists of a cell which is fed with polarized atoms from an atomic beam source (ABS). The polarized atoms scatter from the walls of the storage cell and diffuse via the three parallel tube sections into the vacuum creating a pressure bump at the center (see Fig. 3.2).

The atomic beam source can be configured to deliver either a beam of polarized hydrogen or a beam of polarized deuterium. Deuterium differs from hydrogen in that its nucleus contains an additional neutron. In both cases, the starting point is the molecular gas H_2 or D_2. As the target was designed to run in an underground area where special safety measures apply we used a commercial electrolytic unit for producing the highly-explosive H_2 or D_2 gas without high pressure cylinders.

The first step is to make a cold atomic beam of neutral hydrogen or deuterium atoms. This primary beam is produced by a radio frequency discharge of the molecular gas in a quartz tube for dissociation into single atoms, followed by expansion into the vacuum via a cold nozzle at about 100 K in temperature. The beam is collimated using a conical skimmer. A powerful vacuum pumping system removes the unwanted gas flow to enable free expansion of the nozzle beam at high intensities. Such beams are called "cold" as their relative motion is slow, resulting in a small angular spread, ideal for feeding a storage cell via the narrow feed tube.

Next, the neutral atomic beam (of hydrogen or deuterium) is sent through a system of magnetic fields polarizing the atoms, i.e. the atom spins are predominantly quantized in space. This magnetic field system employs the same principle as that used by Stern and Gerlach in their famous experiment in 1922 in Frankfurt, described in Section 2.1. Instead of a strongly-inhomogenous dipole magnet (two poles), six-pole magnets are used with three North and three

Fig. 3.10: Test of the FILTEX target prior to installation into the TSR. The beam [right to left] of 34.5 MeV α-particles from the accelerator traverses the egg-shaped target chamber (TC) with the storage cell on its axis. The TC is pumped by a big cryogenic pump with charcoal-covered panels at a temperature of 10 K. Two differential pumping stages on both sides limit the gas flow into the experiment. The ABS injecting into the target cell is located behind the TC and partly visible, only. One of the rectangular detector boxes of the TC can be seen in the foreground, allowing for measurement of a left-right asymmetry from which the target polarization could be deduced (*Source*: MPI-K Heidelberg).

South poles and zero field on axis introduced by Nobel Prize winner Wolfgang Paul, acting as lenses on atoms with a magnetic moment. One substate (spin-up) is focused into the target's feed tube, the other one defocused and thus suppressed.

To enhance and switch the nuclear spin polarization of the neutral atoms, RF radiation in a magnetic field of suitable frequency is arranged to cause quantum transitions that result in the nuclear spins being predominantly in one direction. These RF transitions following the adiabatic passage method can reach 100% efficiency. Finally, within the atom in a strong magnetic field, for hydrogen, two nuclear substates of the proton with spin up or down can be populated, and for deuterium — spin one — three nuclear sub-states with spin up, perpendicular and down.

The atomic beam with nuclear polarization now enters the feed tube of the storage cell. The polarized atoms scatter randomly from the coated walls, build up the triangular density of Fig. 3.2, and after exiting the ends of the storage cell they are pumped away by the differential pumping system. To achieve the maximum target density, the straight (beam) tube should be as narrow as possible in accordance with good life time of the stored (electron) beam and low background due to scattering of halo particles.

At the end of the 1980s, ion beams at the Heidelberg Test Storage Ring (TSR) had been commissioned and were used for experiments. The TSR has four straight sections for injection, electron cooling, experiment and acceleration, an excellent vacuum system and a high acceptance, ideal for achieving long storage times important for the spin filtering experiment.

In 1992, the FILTEX target was installed in the TSR. The target properties were investigated using a stored 27 MeV α-particle beam. Clean spectra with high statistics were measured in a couple of minutes, and the dependence of the density (up to $1.0 \times 10^{14}/cm^2$) and the polarization (up to 0.8) on the cell temperature could be straightforwardly studied. Finally, the effect of spin filtering, i.e. the build-up of beam polarization by spin-dependent attenuation in the target, could be successfully demonstrated (Fig. 3.11).

Starting in 1988, there was a parallel design study on the lay-out of a hydrogen/deuterium target for the HERA 30 GeV electron ring. The broad program of development included:

(1) **Design of a storage cell target for the 30 GeV electron beam:** Important issues were: (i) to avoid electromagnetic radiation generated by the pulsed beam by surrounding the beam with closed conducting surfaces (known as *wake field suppressors*), (ii) a storage cell from thin cell walls of high-purity aluminum for reduced scattering and cooling to temperatures below 100 K, and (iii) a system of tungsten (W) collimators to

Fig. 3.11:   The FILTEX target installed in the TSR; in front: Dmitri Toporkov (right), with Frank Rathmann and Erhard Steffens (left) (*Source*: MPI-K Heidelberg).

shield the cell from X-rays, the synchrotron radiation explained in Section 3.6.

(2) **Test of dry-film as a suitable coating for the inner surface of the cell:** A test stand served to study the depolarization caused by wall collisions in a broad range of temperature and magnetic field, using a test cell with on average three thousand wall collisions. The Water effect, ice coating on the storage cell, which significantly improved the target performance was discovered.

(3) **Analysis of the polarization of the target gas:** A prototype of the Breit–Rabi Polarimeter (BRP) was built and its suitability for the precise measurement of the polarization of atoms demonstrated.

(4) Analysis and numerical studies of depolarization of the target gas by the beam and choice of the optimum guide field for H and D atoms.

A crucial issue for high target polarization was a suitable coating of the metal storage cell. During the years 1988–1989, an important validation of a storage cell with dry-film coating in an electron machine was achieved at the Budker Institute of Nuclear Physics (BINP) at Novosibirsk. The coated cell was installed in the 2 GeV/200 mA VEPP-3 electron storage ring by a BINP-ANL collaboration. The cell was fed by the deuterium ABS of the existing polarized jet target. Stable polarization over several months was observed, indicating the suitability of dry-film as coating material in the rough environment of an electron ring.

3.7   Polarized Electrons in the HERA Electron Ring

Electrons (and their anti-particle, positrons) are point-like particles with low mass, about $1/1836$ of the proton mass m_p. As a result, electron beams of energy E_e are typically highly relativistic, i.e. they have a very high Lorentz-$\gamma \cong E_e/m_e c^2$. For the HERA electrons at 27.5 GeV, γ_e is 53,800. For comparison, the 7 TeV LHC proton beam has γ of 7,460, about 7× lower than γ_e. There are two effects particularly important for ultra-relativistic beams of charged particles circulating on a horizontal closed orbit, which scale with the 4th power of γ, and are thus much stronger for electrons:

(1) **Emission of X-rays, so-called synchrotron radiation:** This leads to strong energy losses which increase with the fourth power of the beam energy. For the 27.5-GeV electrons in the HERA electron storage ring, the energy loss of an electron per turn is 0.3% of their full energy. This energy loss has to be continuously replaced by means of superconducting cavities (see Fig. 3.12). This uses a large fraction of the total energy consumption of HERA-e.

(2) **Slow build-up of vertical electron polarization:** This effect was predicted by the Russian theorists Sokolov and Ternov in the 1960s and observed shortly thereafter. It can be understood — according to J. D. Jackson (1976) — by transforming the magnetic field of the bending magnets — about 0.2 T at

Fig. 3.12:   The superconducting HERA RF cavities for replacing the energy of
the stored electron beam lost via synchrotron radiation (of the order of 3 MW,
depending on the beam current) (*Source*: DESY).

HERA-e — into the electron rest frame, a reference frame moving
nearly at the speed of light. This increases the value of the
magnetic field experienced by the electron by the ratio of the
energy to the rest mass — the factor $\gamma_e = 53,800$ introduced
above. The resulting field is in the order of 10,000 T for HERA-e.
The unit 1 Tesla of the magnetic field is about the strength within
fully-magnetized iron, which is quite strong. Therefore, in their
rest-frame the electron spins move slowly into the orientation
of lower energy, for electrons anti-parallel to B, for positrons
parallel, emitting the so-called *spin-flip* synchrotron-radiation.
Because of the different energies of the two spin states parallel
and anti-parallel to the direction of the magnetic field, the flip
rates are different, resulting in a rising polarization. In an ideal
ring, the polarization increases towards a limiting value of 92.4%
with a definite time constant. For HERA-e at the working energy
of 27.5 GeV, one gets a build-up time of 38.6 min. After injecting

a new, unpolarized, electron beam and acceleration to 27.5 GeV, one had to wait for about an hour before a polarized run could start (see Fig. 3.12).

It should be noted that the values of both the polarization and rise time are affected by the machine settings and by nearby depolarizing resonances. An example is the choice of the working energy 27.5 GeV which corresponds to the situation that the number of precessions of the horizontal spin component per turn is just half-integer. Here, the so-called spin tune equals to 62.5, and is the optimum setting of the beam energy for minimum depolarization. The spin tune is related to the anomalous magnetic moment of the electron, a fundamental parameter which can be calculated in Quantum Electrodynamics (QED) and compared to precision experiments.

• **Compton polarimeter:** To determine the scattering asymmetry, the electron polarization of the stored beam must be measured. This is done by back scattering the photons from a circularly polarized laser beam with the electron beam, known as Compton back scattering. In the case of vertically polarized electrons, an up-down asymmetry of the back-scattered photons occurs, from which the beam polarization can be determined.

The transverse electron polarimeter was an integral aspect of the HERA project to enable the study and optimization of the vertical electron polarization produced by the Sokolov–Ternov effect. The basic structure is shown in Fig. 3.13. The interaction point (IP) between electron and laser beams was in the straight West section, about 170 m downstream of the laser lab in the HERA West hall. The photons from an argon ion laser with switchable circular polarization were sent via an evacuated tube to the IP and crossed under a small angle with the electron beam. Back-scattered photons were detected by a tungsten-plastic sandwich calorimeter, developed in the group of Robert Klanner (DESY), which was divided vertically into two

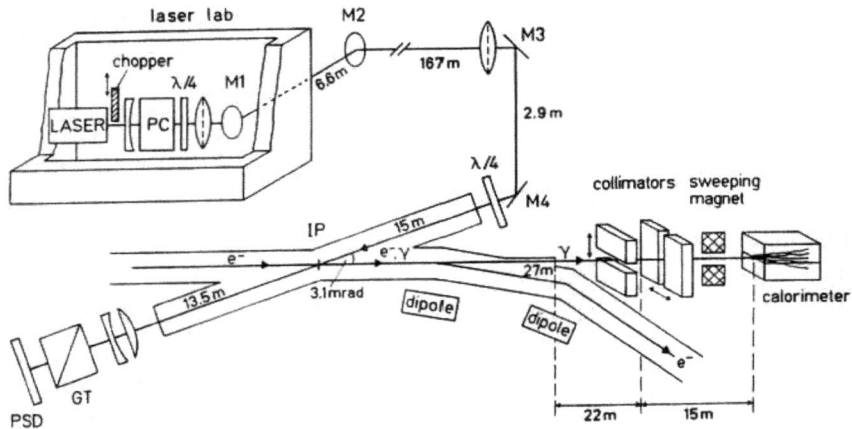

Fig. 3.13:   The HERMES transverse electron polarimeter (TPol) located in the West hall and the corresponding straight section.

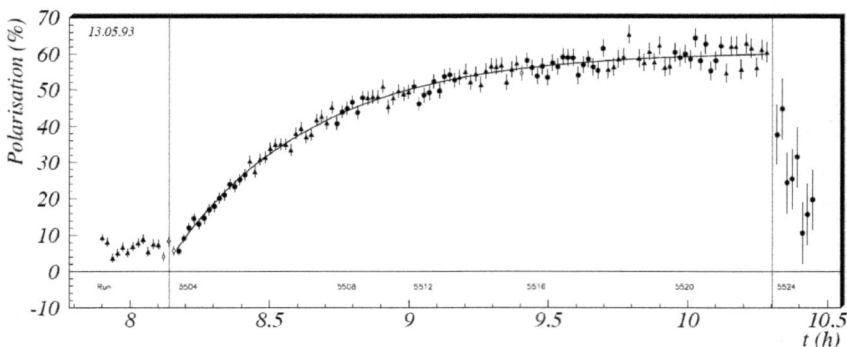

Fig. 3.14:   A rise-time measurement of the vertical polarization by the transverse polarimeter.[9] The asymptotic value P_∞ of the fit is 61%. At $t = 10.3$ h, a solenoid providing a longitudinal B-field was powered, slowly destroying the beam polarization.

halves, allowing for the measurement of the up-down asymmetry of the photon rate.

A measurement of the rise time of electron polarization in HERA is shown in Fig. 3.14, demonstrating a steady state polarization of 61%, well above the required 50%. The highest polarization ever measured was 67%.

Chapter 4

Realizing the HERMES Experiment

4.1 1989–1992: HERMES is Realized

With the arrival of van den Brand at MIT, the internal ^{3}He target development there really got underway. Van den Brand had been an excellent vacuum engineer before deciding to become a physicist. Together with James Kelsey and Ernest Ihloff, two first-rate Bates engineers, both the development of a polarized ^{3}He internal gas target and the design of the target system for the HERA experiment advanced with pace.

It was recognized that it was a priority to find a suitable name for the new internal gas target experiment at HERA. Costas Papanicolas, a native of Cyprus and then a faculty member at the University of Illinois at Urbana-Champaign, named the experiment. He recalls:

> *I vividly remember the collaboration meeting at DESY — I remember that at the end of the meeting Richard Milner raised the issue of the name and offered the prize of a bottle of D.O.M. Perignon. We thought that given "HERA" and "ZEUS" a Greek mythological hero might provide a convenient and catchy name. "HERCULES" (I forget what it stood for) was proposed and I suggested HERMES making also the point that Hercules although very strong he was not regarded highly for his intellect and and actually he was forced to leave the Argonaughts on their expedition to get back the golden fleece. The Argonne people (the "Argonnenauts" as they called themselves) liked the argument. The DESY people liked "HERMES" as Hermes is also the god of Trade and Hamburg*

flourished because of trade. HERMES at the end won and I got
the bottle of D.O.M. Perignon. I opened it when I got elected as
a Professor of Nuclear and Particle Physics at the University of
Athens in 1989, a post I wanted very much at the time.

Throughout the year 1989, the HERMES collaboration worked intensively to develop the proposal. It was agreed to propose the experiment based on the well-established atomic beam source technology for delivering the polarized hydrogen and deuterium atoms. While the laser-driven technology was promising, it had not reached the maturity necessary for HERMES. The high-energy HERA proton beam, which was irrelevant for HERMES, would coast through the experiment and be shielded from the effects of the HERMES magnetic spectrometer with an iron plate with cylindrical bores for the beam tubes. The initial proposal to carry out the HERMES experiment was completed at the end of 1989, is dated as 2 January 1990 and was submitted to DESY on 8 January 1990. The proposal had 16 founding institutions listed in alphabetical order as: Argonne National Laboratory, Caltech, MPI-Heidelberg, University of Illinois at Urbana-Champaign, Los Alamos National Laboratory, University of Wisconsin-Madison, University of Marburg, MIT, New Mexico State University, University of Munich, Stanford University, University of Torino, three institutions from Canada: University of Alberta, Simon Fraser University, and TRIUMF, and the College of William and Mary. The co-spokesmen were Richard Milner and Klaus Rith and the HERMES proposal was presented to the DESY PRC in April 1990.

The major international triennial meeting in subatomic physics, the 12th Particles and Nuclei International Conference (PANIC), took place at MIT on 25–29 June 1990. Discussion of the spin structure of the nucleon was a major aspect of the meeting and the possible funding for the nascent HERMES experiment was being considered. A first contact with Department of Energy (DOE) representatives was unsuccessful.

Later in 1990, the DESY PRC approved the HERMES experiment conditional upon the demonstration of the 50% electron polarization in HERA. While this was a significant milestone, it greatly incentivized the HERMES collaboration to work on realizing the HERA polarization. While HERA had been built to achieve electron polarization, the two major collider experiments ZEUS and H1 were focused on commissioning their large detectors in 1990–1991.

A transverse electron polarimeter located in the West hall was under construction, including a laser system and a 150-m long transport system to the interaction point (see Compton polarimeter Section 3.7). A particularly intensive period was summer 1991, when many HERMES collaborators spent extensive time in the HERA East Hall. No polarization was seen and the studies continued into the fall. On 24 November 1991, a DESY Telegramm, shown in Fig. 4.1, announced that 8% transverse electron polarization had been measured in HERA. By August 1992, over 50% electron polarization had been firmly established. Klaus Rith communicated this to Richard Milner in a fax (Fig. 4.2).

HERMES received full scientific approval, conditional upon realization of the necessary funding, by DESY on 9 October 1992 in a fax from DESY Director Volker Soergel to Richard Milner and Klaus Rith (Fig. 4.3).

In North America, efforts to secure funding to realize HERMES were intensive in the years 1992–1993. A strong Canadian collaboration led by Otto Hausser at TRIUMF took responsibility for the transition radiation detector. In the US, two lepton scattering facilities were under construction, namely the CEBAF accelerator and the Bates South-Hall Ring. Thus, securing funds for HERMES was a challenge. A proposal to the US funding agencies was submitted in February 1991. At MIT, both Robert Redwine, a HERMES collaborator and Director of the Laboratory for Nuclear Science, and Ernest Moniz, Director of the Bates user facility, were active in seeking US funding for HERMES. In 1993, Milner and van den Brand, who had moved to a faculty position at the University of Wisconsin-Madison, were tempted by the offer to move to tenured faculty positions at the University of Illinois. Milner decided to

DESY TELEGRAMM

vom 24. November 1991

Erste Messung
von Polarisation der Elektronen
in HERA

Letzte Woche wurde in HERA zum ersten Mal die Polarisation von Elektronen, die Ausrichtung ihrer "Spins", beobachtet. Im Bereich des geraden Abschnitts HERA-West wurde dazu ein Laserstrahl auf die umlaufenden Elektronen gerichtet, und es wurden die an den Elektronen zurückgestreuten Photonen nachgewiesen. Der Laserstrahl war im Wechsel (90mal in der Sekunde) links und rechts polarisiert. Bei einer Strahlenergie von 26,67 GeV wurde auf diese Weise ein Polarisationsgrad der Elektronen von etwa 8% gemessen. Durch die Veränderung der Beschleunigungsspannung in HERA konnte ihre Polarisation gezielt und reproduzierbar variiert werden. Eine in 10MeV-Energieschritten durchgeführte Messung zeigt Strukturen, die von Depolarisationsresonanzen herrühren.

Elektronen besitzen die Eigenschaft kleiner Kreisel, sie haben einen "Eigendrehimpuls" oder "Spin". In der Teilchenphysik gibt es einige Fragestellungen, die nur mit solchen "polarisierten" Elektronen untersucht werden können.

Fig. 4.1: Announcement from DESY on 24 November 1991 of 8% transverse electron polarization in HERA.

stay at MIT but van den Brand moved to the Free University of Amsterdam and the NIKHEF laboratory. In May 1991, a paper by Frank Close and Richard Milner motivating the detection of the mesons (pions and kaons) coincident with the scattered lepton as a means to further probing the origin of the proton's spin was submitted for publication. Together with similar considerations by European theorists, the scientific case was being made to include the capability for hadron particle identification in the HERMES spectrometer.

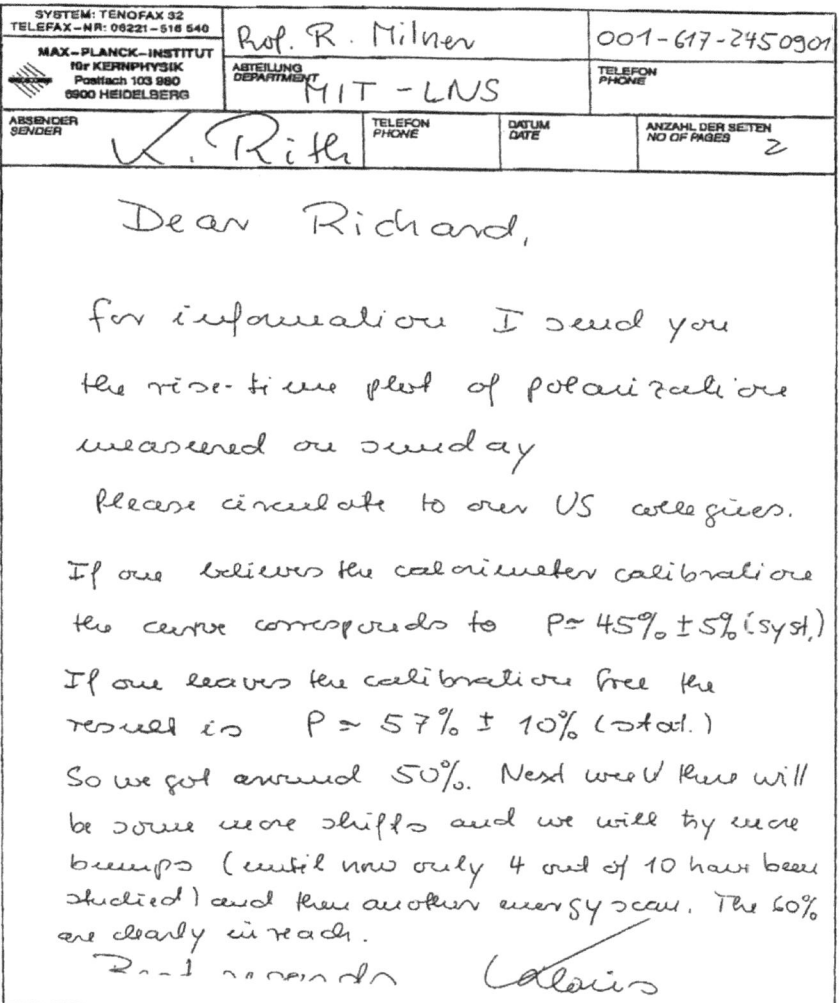

Fig. 4.2: 50% polarization in HERA achieved.

4.2 Funding the HERMES Experiment

With the origination and initial leadership of HERMES coming from Germany and the US, the first approaches to funding agencies took place in those two countries. At the 1990 PANIC conference at

MIT, Milner recalls a discussion with David Hendrie, Director of the US Department of Energy Office of Nuclear Physics, outside of the Kresge auditorium during a coffee break. Hendrie's characterization of HERMES as "an NSF experiment" was an indication that DOE needed some convincing to fund it. In truth, DOE Nuclear Physics had a full plate at that time with both large US flagship facilities, CEBAF and RHIC, under construction. In addition, construction of the MIT-Bates South Hall Ring was getting underway. After the HERMES experiment was granted conditional approval in 1990, serious engagement with the funding agencies worldwide got underway. The HERMES Technical Design Report (TDR) from July 1993 specified that the total cost of the experiment was DM 24 million.

On 18 January 1993 the proposal requesting support for the seven US groups of the HERMES collaboration (Argonne, Caltech, U. Colorado, U. illinois at Urbana-Champaign, U. Wisconsin-Madison, MIT and New Mexico State U.) was submitted to the DOE and the NSF. $3,710,000 was requested over the fiscal years 1993 and 1994. In the US, Sherman Fivozinsky at the Department of Energy and Jack Lightbody at the National Science Foundation played important, constructive roles in ensuring that the US groups in HERMES were adequately funded.

In the half year following Soergel's letter of Fig. 4.3, the various group leaders of the HERMES institutes in Europe and N. America had intense negotiations with their respective funding agencies to secure the funding of their contributions. A visit to the Department of Energy in Washington, DC in early 1993 by Albrecht Wagner was pivotal in securing the funding for the US groups in HERMES.

In 1991, when the former Soviet-Union was in a process of disintegration, Klaus Rith visited two renowned institutes of the former Soviet Union. He recalls:

> *In June or July 1991, I was in Yerevan and Gatchina/Leningrad to negotiate with the institute directors about their respective stakes in HERMES. In the course of my travel to Yerevan, a taxi picked me up at 4 o'clock in the morning from a hotel on the outskirts*

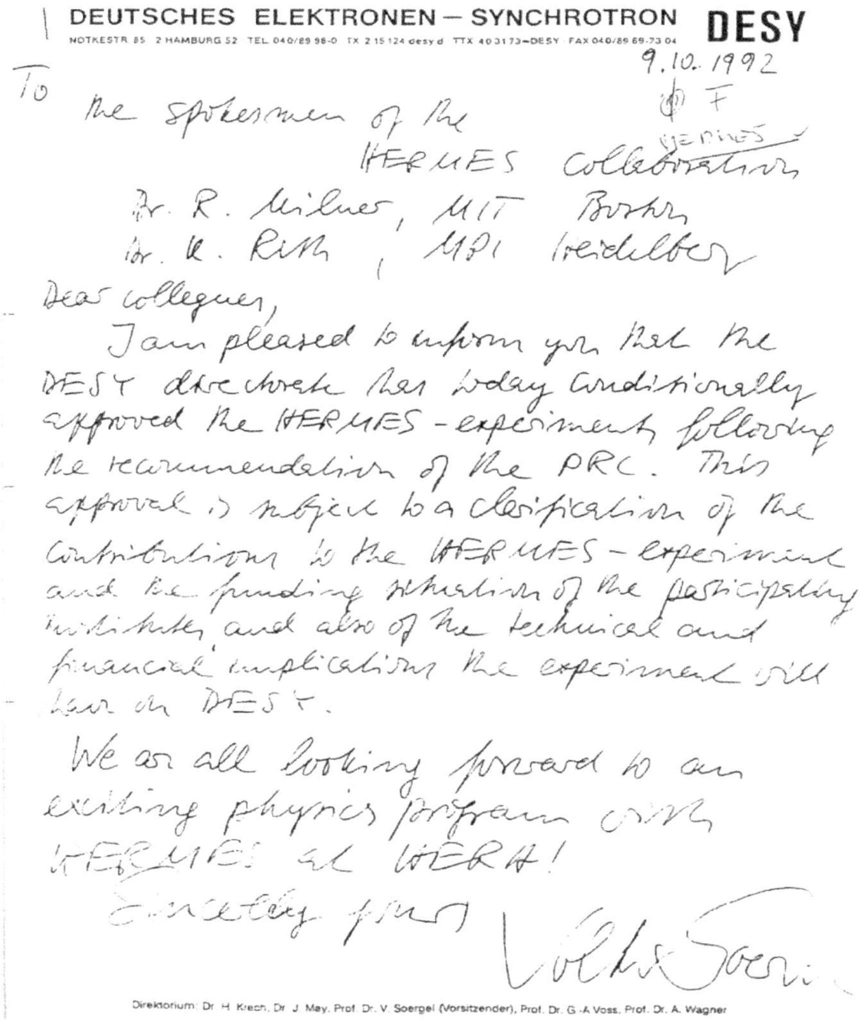

DEUTSCHES ELEKTRONEN — SYNCHROTRON DESY

NOTKESTR 85 2 HAMBURG 52 TEL 040/89 98-0 TX 2 15 124 desy d TTX 40 31 73—DESY FAX 040/89 69-73 04

9. 10. 1992

To

the spokesmen of the
                HERMES collaboration,
Dr. R. Milner, MIT Boston,
Dr. K. Rith, MPI Heidelberg

Dear colleagues,

I am pleased to inform you that the DESY directorate has today conditionally approved the HERMES-experiment following the recommendation of the PRC. This approval is subject to a clarification of the contributions to the HERMES-experiment and the funding situation of the participating institutes, and also of the technical and financial implications the experiment will have on DESY.

We are all looking forward to an exciting physics program with HERMES at HERA!

Sincerely yours,

Volker Soergel

Direktorium: Dr. H. Krech, Dr. J. May, Prof. Dr. V. Soergel (Vorsitzender), Prof. Dr. G.-A. Voss, Prof. Dr. A. Wagner

Fig. 4.3: Fax from DESY Director Volker Soergel to Richard Milner and Klaus Rith communicating the scientific approval of the HERMES experiment, conditional upon realization of the necessary funding, on 9 October 1992.

of Moscow, as my plane from Moscow to Yerevan was supposed to leave early in the morning (from Domodedovo, I think). At the airport I landed in a small section which was apparently reserved for people from the West. Besides me, three American journalists were waiting for the departure, who wanted to report on the conflicts between Armenia and Azerbaijan over Nagorno-Karabakh. People kept coming for other flights, which were then called up after some time and disappeared again. We then tried again and again to find out when our plane would finally leave, but in vain. It only meant 'soon' or 'wait'. I don't remember how we got food, but it must have worked somehow. Late in the evening, it had been dark for a long time, we were finally called and allowed to board a waiting big plane. We were surprised that the four of us were the only passengers, but nothing else happened and the plane door remained open.

Then there was a terrible cloudburst and it poured from buckets. And in this heavy rain the remaining passengers suddenly came running over the runway. Soaking wet Armenians, who had been waiting in another part of the airport, were shooed out of the airport building at the peak of the rain. In a flash it was like a sauna. The plane was now full, the front door was closed and I thought that it was finally going to start. But still nothing happened. Then we were informed that we could not take off because the runway was under water. Finally it stopped raining and the door was opened again. To my astonishment most of the Armenians got up, left the plane and had a kind of picnic on the tarmac. I was well looked after and among other things I was given a chicken leg and I couldn't refuse an Armenian cognac.

At some point it finally started and we landed in Yerevan around three or four o'clock in the morning. The Armenians had disappeared in no time and I stood around quite helplessly as nobody was waiting for me and I had no idea where to go in the middle of the night. (One of Avakian's people had been waiting for me all day, but finally gave it up and left a message with the airport staff which I did not receive.) Fortunately there were still the three journalists who took pity on me and took me in their taxi to the city and their hotel, which was of course completely booked out. The night porter took pity on me and offered me a couch in the foyer so that I could sleep a little under his watchful eyes without being robbed, for which, according to him, the danger was very great.

In the morning, I had them call the institute again and again, where at about 9 o'clock someone finally answered the phone and picked me up. After a short sleep in the guesthouse I gave my HERMES lecture in the afternoon and then concluded an agreement with Robert Avakian and the director of the institute about their participation in HERMES: instead of a financial contribution, which was not possible, the supervision of the production of the lead glass blocks for the electromagnetic calorimeter and the measurement of their properties. (The blocks were later paid for by NIKHEF).

A few days later, the night train from Moscow left for Saint Petersburg (then still Leningrad). This part of the journey was less adventurous, but nevertheless very interesting. I especially remember the following: Stan Belostotski and I had already agreed that PNPI would take over the construction of the magnet. However, the head of the institute declared that this was pure engineering work and that his famous institute should also be given a task for physicists. As a consequence, Stan's team also took over the proportional chambers in the magnet.

There was some imaginative means employed by DESY to make HERMES a reality. The lead glass was manufactured in a Russian factory near Moscow. The Yerevan group played a major role and Harut Avakian recalls he and his father carrying a large amount of cash from Hamburg to Moscow. In Moscow, the money was converted from DM to the local currency and paid to the factory. The lead glass was shipped to Frascati, Italy and to CERN. The first lead glass was tested at CERN and met the required specifications. DESY then gave the goahead for the full production. Somehow the price per block increased over time by a factor of five. In addition, Harut remembers that the factory workers requested supplies of vodka as, apparently, the polishing of the lead glass was of much higher quality after consumption of vodka! In 1993, Harut was hired by Frascati, who played a leading role in the construction of the calorimeter. NIKHEF and Caltech also played a major role in the calorimeter realization.

The final and complete approval of the HERMES experiment was granted by DESY on 15 June 1993, shortly before the collaboration meeting in Rome.

4.3   Groups at DESY Relevant for HERMES

4.3.1   *MEA group*

The Maschine und Experimentelle Anlagen (MEA) Group played an important role in the design, construction, installation, commissioning and operation of the HERMES experiment. The group was located in building 1e at DESY close to the HERMES offices. Klaus Sinram from MEA worked closely with Erhard Steffens, the HERMES Technical Coordinator, in the design of the HERMES experiment in the years 1993–1995. Detailed drawings of components, the necessary mechanical and electrical infrastructure and the many iterations to converge to an optimized design were all part of the responsibility of the MEA group.

Yorck Holler from the MEA group was a particularly dedicated, wise and highly-skilled member of the HERMES collaboration, expert in field measurements of accelerator magnets. He worked long hours in his office in Building 1e and was always ready to help with a technical problem, no matter how obscure or challenging it was. When he came close to retirement he was asked to help finding a successor. It turned out to be virtually impossible to cover all the skills he had left open by a single candidate.

Norbert Meyners was another member of the MEA group who made important contributions essential to the success of the transverse and longitudinal polarimeters.

4.3.2   *Machine department with HERA operations crew*

The machine department was responsible for designing and operating the various accelerators of DESY. Particularly important for HERMES were sections for electron machines (G.A. Voss) and proton machines (B. Wiik) and the HERA operations crew, led by Ferdinand Willeke, now a key figure in the BNL EIC team. Among the members were Des Barber (machine theory) and Eliana Gianfelice-Wendt (from Vaccaro's group at Naples, then LEAR machine group, now at Fermilab), who worked on spin in storage rings, helping to tune the

HERA-e ring to perfection, and Bernhard Holzer, who graduated in the machine group at MPI-K, now at CERN. The HERA operations crew was always willing to tune the beams, with never-ending patience and sometimes with minimal effect only, which then led them to a new island of higher polarization.

4.4 Wagner Meetings

Albrecht Wagner was appointed Research Director at DESY in 1991 and he played a key role in overseeing the approval, funding, construction, installation, commissioning and data taking phases of the HERMES experiment. The monthly *Wagner meetings*, which began in fall 1993, brought together the HERMES Group Leaders, the HERMES collaboration leadership and DESY technical staff to coordinate and manage the many parallel activities (Fig. 4.4). Many significant challenges, both technical and human, were analyzed, confronted and overcome in a calm and effective manner in the

Fig. 4.4:   Albrecht Wagner, DESY Research Director (1991–1999) and DESY Director (1999–2009). While at DESY, he oversaw the full lifespan of the operation of the HERMES experiment (*Source*: DESY).

monthly discussions led by Wagner. The agenda of the first meeting in November 1993 included: the test experiment, the machine coordination meeting, the proton beam dump, the main experiment and Monte Carlo simulation. At the March 1994 meeting, the notes included:

"The final success of the test experiment installation was the result of great effort by too few people. This approach is not appropriate for the preparation of the main experiment. The risk of mistakes and injury due to fatigue would be too high. It is necessary to develop a plan for improving the situation with regard to on-site manpower."

The September 1994 meeting was particularly significant as the collaboration made the decision to install the polarized ^{3}He target for the initial running of the experiment. Wagner's notes from that meeting state:

"Report from test experiment: much has been learned. After extensive tuning and testing, it has been determined that the detectors and reconstruction program work in the HERA environment. Also, the synchrotron radiation tuning seems to be working as designed. The time lines are:

• 17 October 1994: beginning of detector installation
• February 95: scheduled for detector roll-in"

The Wagner meetings were essential to the success of the HERMES experiment and ended in fall 1998. Not surprisingly, Wagner was promoted to DESY Director in 1999, following the untimely accidental death of Bjørn Wiik.

4.5   Design and Construction of the HERMES Experiment in the HERA East Hall

Steffens — a personal remark: The final approval of HERMES in July 1993 changed Steffens' life completely for the next two years. His role as Technical Coordinator required his weekly presence at DESY, and thus frequent trips to Heidelberg-DESY by train, about five hours, plus local traffic. The number of meetings increased strongly to the possible maximum or beyond, which was compatible with a family living near Heidelberg.

A typical week #27/1993 from his diary: Mon — Wed DESY (disc. detector mounts, gas meeting, meeting with L. Kramer/MIT about vacuum system of ^{3}He target); Thu–Sat MPI-K Zelenski, visitor BNL, talk, Group excercise at Univ. (as Coordinator for 10 groups, including his own group); Sunday at home, evening: travel to DESY...

One year later in 1994, same week #27: Sun. previous week: travel to DESY; Mon. — Thu.: Experiment Coordination, small meeting with Wagner, Group leaders, Wagner-meeting, Onsite-meeting, Techn. meet., Target meet., meeting about Cryosystem; Fri. MPI-K, disc. with local target group, admin. business; weekend at home, Sunday evening: travel to DESY...

4.5.1 *Infrastructure at the HERA east hall*

Once the HERMES experiment was conditionally approved in October 1992, both the collaboration and DESY established the goal that the experiment should begin data taking in 1995. This meant that detector installation should begin in late 1994. Further, it was agreed that successful operation of a test of an (unpolarized) internal gas target should take place in 1994, which required installation in the 1993/1994 shutdown.

The tasks were divided between the host institute DESY and the HERMES-Collaboration in the usual way:

- DESY: Infrastructure and operation of the accelerator;
- HERMES: Components of the experiment (targets, detectors, data-acquisition), running and analysis.

HERMES was located in the huge underground HERA East-Hall which was completely empty when the project started. In addition, plenty of space was available in the 8-storey underground appendix to the hall with stair case and elevator. This was used as lab space for the different groups for working on their components. The design of the infrastructure started in early 1993 and was coordinated by Yorck Holler (MEA-DESY) and Erhard Steffens (HERMES).

The Z-Division of DESY was responsible for "Central Installations and Facilities". Of special importance for HERMES were:

- MEA Group (machine and experimental facilities), led by Klaus Sinram (both Sinram and Holler were members of HERMES);
- Mechanical workshop including the Design Office.

In addition, the Z-Division comprised the electronic workshop, computer center, administration and other services.

The Design Office was particularly important for defining the infrastructure necessary to mount and supply all the components of the experiment depending on their needs. HERMES was the first experiment at DESY where 3D design tools were applied. The know-how came with a small private office with two young engineers who had just finished the design of a complicated mixing facility with many tube lines in three-dimensional for the Beiersdorf Company ('Nivea skin cream'). Their new software allowed for a precise design of components close to each other with arbitrary orientation in space. Without such tools, the components had to be drawn with several cuts or projections to be sure that neighboring parts do not overlap.

The basic structure was laid out in the Technical Design Report (TDR, July 1993). The experiment was located on the Experiment Platform (EP) movable on rails, and rigidly connected to another movable platform carrying the two-storey Electronic Trailer (ET) for electronics and gas systems. There were three positions of platform and trailer:

(1) Beam position: EP inside shielding in beam position, ET outside;
(2) Tunnel access for Tram: EP moved to the side within shielding;
(3) Parking position: Both ET and EP outside the shielding.

Position (2) allowed quick access to the tunnel in case of a failure of a heavy machine component, e.g. a magnet.

Figure 4.5 shows the HERA East Hall in December 1993 when the hall was prepared for the installation of HERMES. The shielding blocks to the hall are removed. The elevated thick concrete floor in the background is at tunnel level. The machine axis runs parallel to

Fig. 4.5: The HERA East Hall in December 1993, see text. The responsible Hall engineers were Gerd Wöbke and Wilhelm Beckhusen (*Source*: DESY).

the backward shielding wall. The machine components around the Interaction Point (IP) were removed to allow the installation of the e- and p-beamlines in its new configuration around the HERMES IP, together with the test experiment (see Fig. 4.6), consisting of (i) the collimator system (in front), (ii) a target chamber with wakefield suppressors and test storage cell with gas inlet, and (iii) prototypes of the different tracking detectors and the calorimeter. Studies performed in the 1994 run with single and colliding beams showed that, when the movable collimator was closed, the background from synchrotron radiation was strongly suppressed. Further, the beam lifetime was not affected by gas introduced into the target cell.

In parallel to the running of the test experiment in 1994, the assembly of the complete experiment commenced down in the hall behind the shielding wall. After the installation of the rail system, the platform for supporting the spectrometer magnet (250 tons) and detectors including scaffolding (150 tons) was delivered and craned down. The magnet had to produce a strong field of about 1.5 Tesla in a large volume. Its function is described in Section 3.5. It was

Fig. 4.6:   Gustav-Adolf Voss, Bjørn Wiik and Gerd Wöbke at the HERMES test experiment in the HERA East Hall in 1994 (*Source*: DESY).

designed and built in Petersburg, Russia. It consisted of numerous elements of mass below the limit of the hall cranes. In Fig. 4.7, the assembled magnet on the experimental platform (EP) is shown, next to the shielding wall.

Also in 1994, many components of the detector systems arrived and were assembled and prepared for installation. An important element of the infrastructure was the cabling between the ET and the experimental platform (EP). All cables and gas or cooling pipes had to be inserted into several flexible cable trays. The dimensions were obtained with the help of the three-dimensional design program. A huge effort went into collecting all the information from the users and to manufacture the cables at the correct length, resulting in a new weekly meeting "Cabling", led by Steffens. Cabling was handled by an experienced external company and took three months, only. In total, about 200 km of cables had to be laid, including 33 km of coaxial cable for the calorimeter alone.

Prior to mounting the detectors, the field distribution of the magnet was measured using a machine borrowed from the GSI

Fig. 4.7:   The Electronic Trailer (ET) and HERMES experiment in October 1994. The platform of the experiment with spectrometer magnet is visible on the right, next to the shielding wall. The ET's (left) lower level is for the electronics in water-cooled racks, the upper level for the gas systems. In the foreground, the HERA-tram is visible. This was designed for removal or installation of heavy machine components like magnets (*Source*: DESY).

laboratory in Darmstadt. The measured three-dimensional-field map was required to compute the bending power of the magnet for tracks of charged particles emerging from the target. The reconstruction of the charged particle tracks was an essential aspect of the data analysis — see the HERMES event displayed in Fig. 4.13 and the discussion of particle tracking in the following Section 4.5.2.

4.5.2   *Preparation and installation of the detector components*

The construction and installation of the HERMES experiment on a rapid timescale by the institutes of the HERMES collaboration was an outstanding achievement. In the twenty months between the final approval and the closing of the tunnel, the HERMES collaboration ordered all the components, constructed and tested the detector elements and assembled the complete HERMES spectrometer. This required extraordinary coordination across the institutes. HERMES Deputy Spokesman Mike Vetterli played a central leadership role in this great success.

The HERMES Spokesperson during this period, Jo van den Brand, recalls:

> *Most of the components of the HERMES experiment were installed in the HERA East Hall in the course of 1994. That year I was appointed Spokesperson of the HERMES Collaboration. As a young nuclear physicist I was eager to work at a renowned particle physics laboratory. Firstly, the 250 ton spectrometer magnet was build-up in record time on the experimental platform and put into operation on July 21, 1994. The installation of various detectors started in October. At the end of 1994 most detectors behind the magnet were installed: the drift chambers, hodoscopes and TRD. As one of the biggest components the completely instrumented electromagnetic calorimeter (25 ton total weight) found its destiny on the HERMES experimental platform on December 13, 1994. NIKHEF was the prime responsible for this calorimeter, and I was then also the leader of the calorimeter project. In addition NIKHEF committed to the realization of the vertex detector: a 25,000 channel tracking detector based on MSGC technology.*

The individual detector components were positioned by means of a scaffolding with adjustable mounts, except for the calorimeter which had its own frame. The detectors were produced at the respective home institutions and shipped to DESY where they were prepared for final installation, scheduled for the last 3 months (Nov. 1994–Jan. 1995) before the rolling-in of the complete experiment into beam position. The three Magnet Chambers (MC's) located between the spectrometer magnet poles had to be installed first. As they were hidden behind massive steel plates, the field clamps, they could only be accessed during longer shutdowns. For tracking of the charged particles, gas-filled wire chambers of different types were used, four in front, three within and four behind the spectrometer magnet.

In Fig. 4.8, a vertical cross section of the experiment in its 1996–1997 configuration is shown. The pole of the magnet (blue)

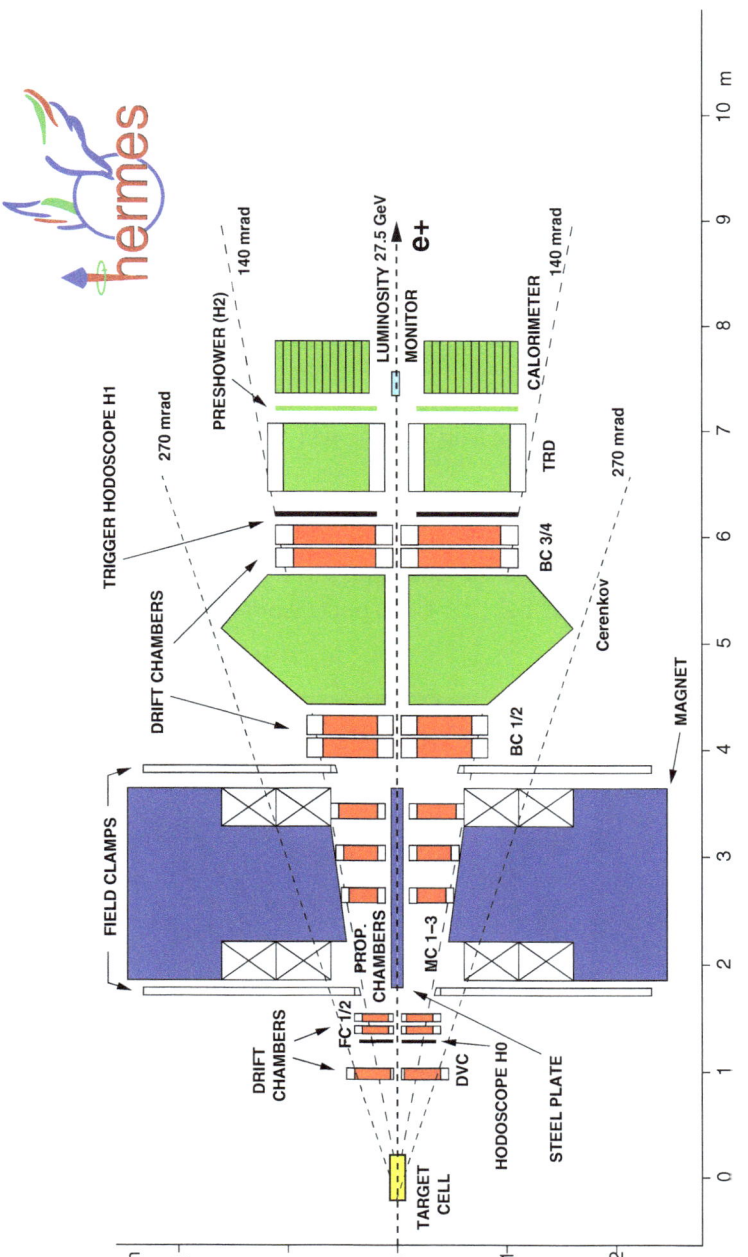

Fig. 4.8: The HERMES spectrometer in its 1996–1997 configuration (for details see text) (*Source:* HERMES Coll.).

is shown together with the steel plate, shielding the beam tubes from the magnetic field in the gap. Additional vertical steel plates (field clamps) serve to confine the field region. The tracking detectors are shown in red, the hodoscopes H0–H1 in black, and the PID (particle identification) in green. The vertical acceptance angles of the experiment (nominal 140 mrad) and the DVC alone (270 mrad) are indicated. The fast hodoscopes based on plastic scintillators were part of the trigger system initiating read-out of the detector. Two gas detector systems, the Cherenkov ($\check{C}$) and the transition-radiation detector (TRD), served to identify the particles (e, π, K, γ, ...), assisted by the hodoscopes and the calorimeter (Figs. 4.9–4.12).

The individual detector components and essential elements for physics analysis are summarized as follows:

- **Drift Chambers for Forward Tracking:** The Vertex Chambers (VC)/NIKHEF, and the Front Chambers (FC)/Colorado provided precise vertices of the tracks emerging from the target, before they entered the magnetic field of the spectrometer magnet. Later, it

Fig. 4.9:   Installation of the Cherenkov detector for PID (*Source*: DESY).

Fig. 4.10: The first row of the HERMES data acquisition (DAQ) in the ET is shown, where all detector signals arrive and are processed.

was found that the performance of the VCs was compromised by their faulty readout electronics, and these chambers were later replaced by the Drift-Vertex Chambers (DVC/Dubna).

- **Magnet Chambers:** The task of the Magnet Chambers (MC)/ Petersburg/Rome was to connect front and backward tracks for a precise measurement of the bending angle and thus the momentum of the scattered particle. They also helped in the detection of short-lived particles like the Λ.
- **Drift Chambers for Backward Tracking:** The Back Chambers (BC 1,2)/Erlangen and (BC 3,4)/Zeuthen completed the momentum measurement from magnetic deflection in the field of

Fig. 4.11:   The transition radiation detector (TRD) for particle identification (PID) in place. From left to right: Robert Openshaw, Andy Miller and Brad Filippone (*Source*: DESY).

the spectrometer magnet. Because of their large width, they had to be built with pre-stressed frames.

- **Cherenkov Detector:** The gas-filled threshold-$\check{C}$/Argonne was part of the particle identification system (PID) and served to separate pions from other hadrons. Later, a second solid low-density radiator made of Glass Foam was added and the detector converted to a Ring-Imaging CHerenkov (RICH).

- **Hodoscopes:** The two hodoscopes (H1 and H2)/Caltech with 42 plastic scintillator paddles, each, were part of the main trigger and, in addition, contributed to PID. Later, more hodoscopes were added to improve background suppression.

- **Transition Radiation Detector:** The TRD/(TRIUMF, Simon Fraser, Alberta) was based on X-ray emission of highly relativistic charges, in particular electrons, when crossing boundaries of media with different electric permittivity ϵ_{rel}. After adding a thin lead curtain to reduce the effect of synchrotron radiation, they reached a high sensitivity for electrons.

Fig. 4.12: Installation of the lead-glass electromagnetic calorimeter. The loaded frame is craned into position. The group in front includes Brad Filippone and Jo van den Brand. The group on the balcony behind includes Wolfgang Korsch, Allison Lung and Michael Düren (*Source*: DESY).

- **Lead-Glass Calorimeter:** The electromagnetic calorime-ter/(Frascati, NIKHEF, Yerevan) consisted of 840 lead-glass blocks $9 \times 9 \times 50$ cm^3 arranged in two halves of 6 tons each in mass. The blocks were read out by 840 separate photo-multiplier tubes. The blocks were suspended by a massive frame designed so that they could be moved apart during injection to avoid radiation damage. The performance of each block and phototube was determined in test beam facilities at DESY and CERN in 1993/1994, and was monitored continuously during data taking with a gain monitoring system. Pion rejection proved to be better than a factor of 100, while the resolution was about 4% at 20 GeV.
- **Luminosity Monitor:** The luminosity monitor/(Erlangen, Lebe-dev) was located downstream of the experiment close to the beam axis behind exit windows, and detected pairs of electrons from the scattering of fast beam electrons from target-electrons at rest. Its event rate was proportional to the product of beam current

and target density, the so-called luminosity (see explanation in Appendix A) of the scattering process.

- **Trigger System:** The trigger system (New Mexico) provided a fast signal for a certain class of events. Example: DIS = deeply-inelastic scattering event, required a coincidence of (H1 + H2 + CALO + HERA-clock).

- **Particle Tracking:** The reconstruction computer program combined the information from all the tracking detectors, the known magnetic field of the HERMES spectrometer and information from other detector components to determine the momentum of the particle. It was developed by Klaus Rith's group at the University of Erlangen and is described in detail in the Ph.D. thesis of Wolfgang Wander. An example of a reconstructed 3-track event is shown in Fig. 4.13. The hits in the various detectors are indicated by circles.

- **Particle Identification (PID):** Information from the Cherenkov and TRD detectors was combined with information from the calorimeter and hodoscopes to construct a probability distribution associated with the identity of each different type of particle. This particle id software was developed by TRIUMF, SFU and Caltech.

- **Data Acquisition and Computing:** The complete suite of HERMES detectors was controlled by computers, triggered when an interesting event was detected in coincidence with the presence of the HERA electron beam, and all the detector information for a given event was read out and carefully stored on a computer disk. Then the entire HERMES detector system was reset and the system prepared for the next triggered event. The data acquisition system and computing was the responsibility of Heidelberg and was led by Walter Brueckner.

4.6   HERMES Running I: 1995–2000 with Longitudinal Target Spin

4.6.1   *Installation and commissioning of the HERMES experiment with the polarized ^{3}He target*

In September 1994, the HERMES collaboration made the decision to install the polarized ^{3}He target for the initial 1995 data taking. This

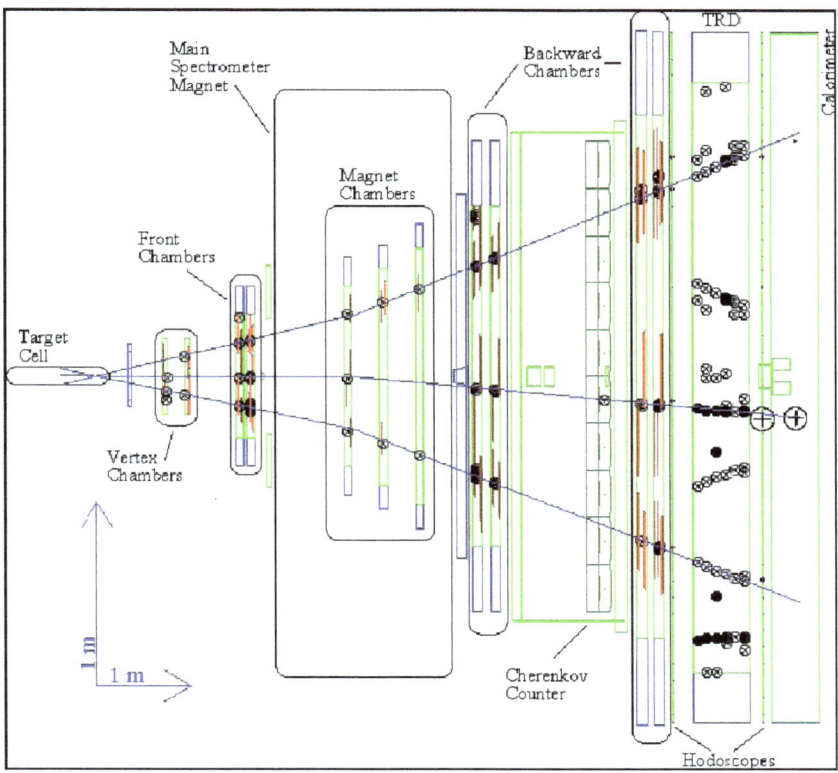

Fig. 4.13: Picture of a reconstructed 3-track event. See Fig. 4.8 for the spectrometer layout. The hits in the various detectors are indicated by circles. The tracks start in the target cell (left) and run straight up to the magnet, where they change direction, and then continue straight to the calorimeter, where they are stopped (*Source*: HERMES Coll.).

was motivated by two main considerations. Scientifically, there was great interest in measuring the neutron spin structure function and SLAC was also planning a run on polarized ^{3}He in 1995. Technically, the polarized ^{3}He target was judged to be less risky compared to the polarized hydrogen/deuterium target. A polarized ^{3}He internal gas target system had previously been built at MIT by Milner's group and successfully used in experiment CE-25 at the Indiana University Cyclotron Facility in 1991, as described in Chapter 3.

As a noble gas, the ^{3}He atom has electron spin zero and a nuclear magnetic moment close to that of the free neutron. In

Fig. 4.14:   The HERMES polarized ^{3}He gas target (*Source*: DESY).

contrast to hydrogen atoms, the spin of ^{3}He atoms slowly follows directional changes of external magnetic fields, which serve to define the polarization direction. In order to preserve the polarization of the gas, all components of the target are arranged inside a large-volume magnetic field, produced by current carrying coils which can be seen in Fig. 4.14.

The polarized ^{3}He target was developed in a common MIT-Caltech effort. At MIT, Stephen Pate, Laird Kramer, Richard Milner and graduate student Dirk DeSchepper together with MIT-LNS engineer Jim Kelsey, designed and constructed the optically pumped polarized ^{3}He source together with a non-magnetic target chamber and complete differential vacuum pumping system. This vacuum system was subsequently used for all HERMES targets until the end of running in June 2007. MIT designer Ron Filosa worked with Jim Kelsey for the Laboratory for Nuclear Science and their office was located in building 44 on Vassar Street, Cambridge. The MIT group worked intensively on the design of the target for the test experiment in 1994 and redoubled their efforts in September that year when the HERMES collaboration made the decision to take data with the polarized helium-3 target in the initial year of running.

Polarized ^{3}He gas is produced by metastability exchange laser optical pumping in a quartz cube and the atoms are then transferred via diffusion through a glass capillary tube into the storage cell, cooled to a temperature of about 30 K inside the target chamber. A laser at wavelength 1.083 μm is used to optically pump from the 2^3S metastable state to the excited 2^3P state in helium. The laser was constructed from a commercial YAG laser, using a lasing medium closer to the desired wavelength and employing intra-cavity tuning for optimal polarization. The class IV laser system was located in an inter-locked hut in the East Hall and the circularly polarized light transmitted to the pumping cell via a series of mirrors and lenses. The quartz glassware was made by the master glass blower Gerhard Finkenbeiner and his staff in Waltham, MA and shipped to DESY. Finkenbeiner was a German master glassblower and unique in being a manufacturer of the glass armonica, a musical instrument invented by Benjamin Franklin in 1761. Finkenbeiner's armonicas have been played at the Metropolitan Opera, in radio commercials and in movies. Mysteriously, in May 1999, Finkenbeiner took off in his single-engine plane and neither he nor the plane have been seen again. G. Finkenbeiner Inc. has continued as a successful glass making business.

Optimal design of the storage cell included minimizing the storage cell transverse dimensions to maximize the target density while keeping it large enough to minimize interactions of beam particles. These interactions included the absorption of synchrotron radiation photons as well as the scattering of beam halo positrons into the detector. A related issue was that the collimator system that shielded the cell and the detector from this radiation had to have an aperture that allowed for some beam steering. The cell had to be positioned to be in the shadow of this collimator system.

The storage cell was a long elliptical cylinder with its central axis positioned along the beam axis. The elliptical shape of the cell mirrored the beam shape, and the size corresponded to the specified clearance for the beam during injection. To increase the target density, the cell was designed to run at cryogenic temperatures down to 15 K. This requirement forced the use of ultra-pure aluminum (99.9999%), which has an excellent thermal conductivity at cryogenic temperatures, being more than 30 times as conductive as commercially pure aluminum in the working temperature range. The storage cell was fabricated from two 200 mm × 550 mm sheets of 125 micron thick ultra-pure aluminum which were formed into half ellipses between two molds. A feed tube along one side was formed to allow injection of the polarized ^{3}He atoms. The two foils were then spot welded together along the sides of the ellipse, along the outer edges of the aluminum sheet and along the sides of the feed tube.

The storage cell was cryogenically cooled to increase the target density without increasing the gas load to the storage ring. This was accomplished by flowing cryogenic helium gas through the storage cell support rails. The cryogenic helium gas was supplied by a dewar located near the target chamber. This dewar was fed by the DESY HERA storage ring cryogenic supply and refilled between data taking periods.

Beam bunches traveling around the storage ring induce a corresponding image charge in the beam pipe which follows the bunch around the ring. If this image charge encounters any discontinuities in the beam pipe, wake fields may be generated. These wake fields can cause energy loss in the beam which translates into heating of the offending components. It can also change the momentum and emittance of the beam, and therefore has to be avoided throughout the accelerator ring. To minimize these effects, the cross section of the beamline is kept constant as much as possible. This was achieved by using so-called *wake field suppressors*: a thin-walled titanium mesh connecting to the target cell with beryllium–copper fingers.

For high polarization in the pumping cell, the transverse magnetic field gradients from the stray field of the HERMES spectrometer had to be actively compensated by a custom-designed magnetic coil system. This was designed and constructed in winter 1994.

At Caltech, important studies using a low energy deuteron beam were carried out to establish the optimal temperature of the target

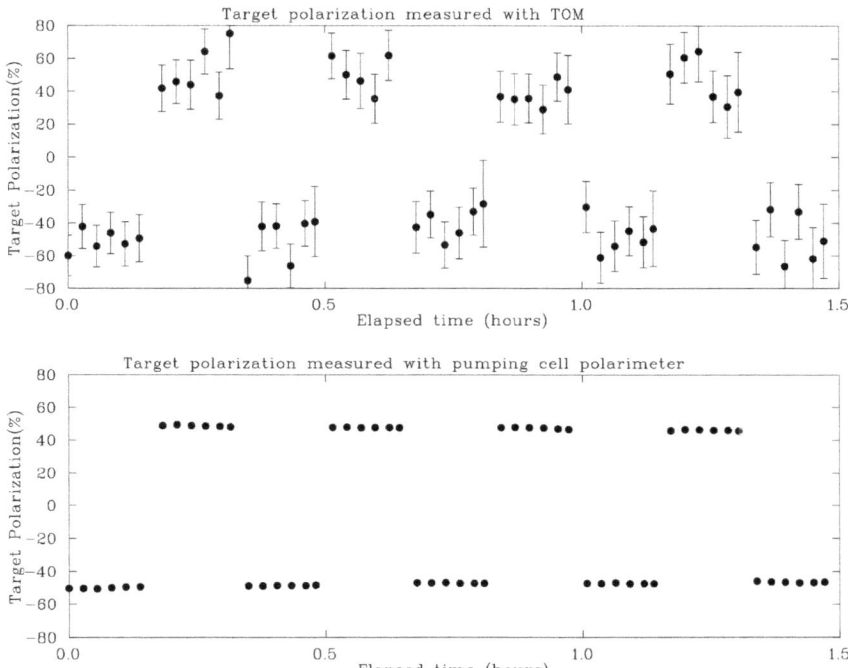

Fig. 4.15: The polarization of the HERMES polarized ^{3}He target vs. time measured with the target optical monitor (upper panel) and the pumping cell polarimeter (lower panel). Each point represents the average of 100 s of data (*Source:* HERMES Coll.).

cell. Further, a target optical monitor was designed and constructed to measure the polarization of light produced by excitation of the polarized ^{3}He gas in the target cell by the 27 GeV HERA electron beam and was cross-calibrated with a measurement of the polarization in the pumping cell. Figure 4.15 shows a comparison of the ^{3}He polarizations measured in the pumping cell (lower panel) and in the target cell (upper panel). The Caltech HERMES target group consisted of Robert Carr, graduate student Andrea Dvoredsky, post-docs Wolfgang Korsch and Mark Pitt, Prof. Bob McKeown and their colleagues.

The HERMES polarized ^{3}He target had a thickness of 1×10^{15} ^{3}He atoms/cm^2, limited by the constraint that it reduce the HERA

Fig. 4.16:   Rolling-in of the HERMES experiment into beam position. The proton and electron beam lines are visible in the foreground (p left and e right) and at the tunnel entrance in the background (*Source*: DESY).

stored beam lifetime by no less than 45 h. The unique gas feed system of the polarized ^{3}He target could also be used to deliver unpolarized nuclear gases, e.g. ^{4}He, neon, krypton, and argon. This was subsequently used to take scattering data on these targets which had important scientific relevance for understanding the heavy-ion data from RHIC.

On 6 February 1995, the detector was moved into beam position (see Figs. 4.16 and 4.17) and the East hall was prepared for the

Fig. 4.17: Side view of the HERMES experiment in beam position (*Source*: DESY).

1995 machine run. On 31 March at 9:00 the Interlock was set for the first time, and HERA started operating, 21 months after final approval of the HERMES experiment. At the September 1995 DESY PRC meeting, the HERMES referee David Saxon reported "that the experiment had entered the measurement phase and the committee should congratulate the experiment." He reported that the HERMES sub-detectors were working well, except for a readout problem in the vertex detectors.

In the fall 1995 data taking period, HERMES acquired more than 5 million events using the polarized ^{3}He target with polarization of both beam and target of about 50%.

Fig. 4.18:   The HERMES longitudinal electron polarimeter (LPol). The photo
shows the view along the motion of the electrons into the tunnel which in the
distance starts to bend to the right. The vertical tube for the transport of the
laser beam from the laser lab upstairs to the inflection point is visible (see bright
spot) where the light is directed head-on to the electron beam. The back-scattered
Compton photons are detected downstream in the tunnel where the bend starts.
For demonstration purpose, the green light beam from an argon-ion Laser is
deflected out of the beam tube by Marc Beckmann (*Source*: DESY).

A new, *longitudinal*, electron polarimeter, built by U. Penn and
Freiburg, was installed in the 1995/1996 shutdown in between the
two spin rotators in the HERA East straight section to reduce
the uncertainty in the determination of the beam polarization (see
Fig. 4.18).

4.6.2   *Installation and running of the hydrogen and deuterium (H/D) target*

The basic requirements of the targets were (i) high density and
polarization, and (ii) determination of the nuclear polarization of
target protons (or deuterons) by the target diagnostics with high

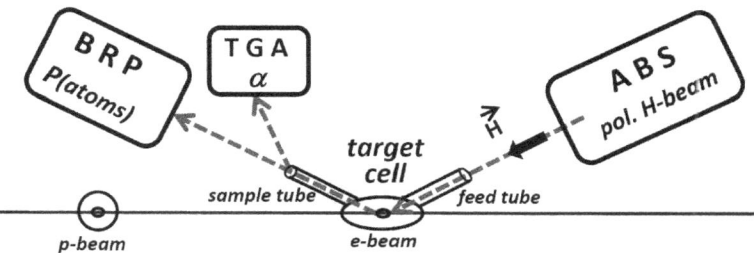

Fig. 4.19: The schematic diagram shows a vertical cross-section of the HERMES H/D target with the main elements ABS, target cell with feed and sample tubes, and diagnostics, consisting of TGA and BRP. The view is towards the spectrometer magnet, in the direction of the electron beam as in the H/D target photo of Fig. 4.20 (*Source*: E. Steffens).

accuracy. This resulted in a complex apparatus shown schematically in Fig. 4.19, and in the photograph of Fig. 4.20. The different components are:

- **Atomic Beam Source (ABS) (Heidelberg, Marburg, Munich):** The polarized beam is produced, starting with an RF-discharge by dissociating hydrogen molecules H_2 into H atoms, forming a beam of cold hydrogen atoms, the spin-up component being focused by strong sextupole magnets (three North and three South poles) into the feed tube of the cell. The spin polarization of the H beam is switched periodically by means of a system of radio-frequency transitions.

- **Target Cell with Chamber and He-cooling (Liverpool, Madison):** A tube of elliptical cross-section from aluminum sheets 75 microns thick coaxial to the 30-GeV stored electron beam, and suspended by two parallel cooling rails with temperature down to 50 K was the fragile heart of the target. The cell was designed and supplied for many years by the Madison group, in particular Thomas Wise, who brought spare cells which he had made by air travel to Hamburg. Together with the target chamber (Liverpool group) with He cooling unit it turned out to be a reliable component of the target. An initial problem was corrected by an improved design where the constant cell cross section was

Fig. 4.20:   The polarized hydrogen/deuterium gas target as installed in HERA 1996–2000. The electron beam line is in the center with the target section removed. The front face of the spectrometer magnet is in the background. The ABS is on the right side, injecting into the target chamber. The target diagnostics on the left is located above the proton beam line (see schematic in Fig. 4.19 for comparison). In 2001, the target was modified by installing the transverse magnet (*Source*: DESY).

complemented by bulges which avoided an inward-bending of the cell wall in case of thermal expansion which happened a few times affecting the stored beam. For the last 6 years of HERMES running, the cells were supplied by Ferrara. The direction of the target spin parallel to the beam axis is defined by a longitudinal magnetic field produced by means of superconducting coils. Its strength was carefully chosen to avoid depolarizing resonances caused by the time dependent magnetic field of the electron beam (see Section 4.7.2).

- **Diagnostics TGA and BRP (Heidelberg):** The diagnostics (see Fig. 4.19) were developed by the Heidelberg group, together with Dmitri Toporkov (Novosibirsk) during his stay in 1991. The sample beam is analyzed by (i) the Target-Gas Analyzer

(TGA), sensitive to the fraction of molecules, and (ii) a polarimeter (Breit–Rabi polarimeter, BRP), essentially an inverted ABS with the detection of hydrogen atoms at the end, which determines the nuclear polarization of the target atoms.

The installation took place during the Winter shutdown 1995–1996, followed by the running-in. A critical issue was the coating of the target cell in the difficult environment at HERA-East where typically several hundred Watts of synchrotron radiation traverse the collimator system and cell. A dry-film-coated cell had been used in the 1994 test experiment (see Fig. 4.6), and afterwards showed clear signs of radiation damage. From laboratory studies, described in the thesis of Bernd Braun, it was known that running the cell at temperatures below 100 K might cure the target cell surface by the addition of an accumulated ice layer. During initial operation with a cold target cell and electron beam, a rising fraction of atomic hydrogen surviving recombination was observed by means of the TGA, confirming earlier measurements. The fraction of hydrogen atoms in molecules during the first period with longitudinal target spin was about 5%, and was stable over longer periods without electron beam losses in the vicinity of the target. In case of such losses, the fraction of atoms went down, probably due to damage of the ice layer, but recovered within a couple of days.

It should be noted that the recovery-effect of the cell surface without action from outside, e.g. venting, was mandatory for the success of the H/D target! We might call it the *Braun effect*, as it was first seen by Bernd Braun. In retrospect, this effect also explains why the measured degree of dissociation in storage cells at temperatures below 100 K was often high, independent of the surface material.

• **Factors Determining the Target Polarization Seen by the Beam:** The electron beam traverses the 40 cm long gas target. Important for the measurement are: the projected (areal) density of target protons p (p/cm^2) and the average spin polarization of the protons. The diagnostics measures the molecular fraction α and the polarization P of the sample beam, i.e. that of the cell center. When the atoms diffuse outwards via the three openings of the T-shaped cell (see Fig. 3.1), they (i) perform wall collisions — in the average a couple of hundred — and may gradually depolarize, (ii) they may recombine to H_2 with a partner H atom at the wall resulting in partly depolarized molecules, (iii) or they may exchange their spin with another atom during a rather unlikely collision in the volume, losing nuclear polarization. This results in a profile along the beam axis z, with dependences $P(z)$ and $\alpha(z)$. In the simplest case, with no such processes, one would expect constant values, i.e. flat profiles. The stronger relaxation is, the steeper the profiles are. They are described by the so-called Sampling Corrections (SC), see below.

In order to separate the different mechanisms one has to vary the conditions. Changing the density of atoms will affect Spin Exchange, changing the cell temperature may affect the wall properties, etc. By means of so-called Monte Carlo (MC) calculations, the diffusion of atoms and/or molecules is tracked during their motion through the cell until their leave. This motion is particularly simple in the Molecular Flow Regime, dominated by wall collisions, only, which can be assumed as fast adsorption, followed by re-emission according to a $\cos(\theta)$-law, θ being the angle with the normal to the cell wall. This implies that the motion is random and depends on the geometry of the walls, only. Mostly likely is re-emission perpendicular to the wall. For a narrow long tube, it takes many (several hundred) wall collisions to leave the tube by its opening, resulting in a high target density of the storage cell.

• **Sampling Corrections (SC):** These are the corrections to quantities measured at the center of the target cell in order to obtain the quantity at a specific position along the cell axis. These corrections are small or negligible e.g. for a high-quality coating. The precision with which those corrections can be determined by measurements and simulations, is of highest importance for the experiment, as they enter directly into the total measurement error. The better the cell walls are prepared, the smaller these corrections its opening, resulting in a high target density of the storage cell.

The first 2 years of running the H/D target were devoted to running-in and measurements with longitudinally polarized hydrogen (H_{long}). In the 1997 run, the average polarization was 0.851 ± 0.033, and the areal density was 0.7×10^{14} nucleons/cm^2.

In the years 1998–2000, the target was run with deuterium. The deuterium nucleus is known as the *deuteron*, which has nuclear spin-1, i.e. there are three possible spin directions with respect to a magnetic field: parallel, perpendicular and anti-parallel. A complete set of deuteron spin observables requires more measurements than for the proton case. The deuteron target spin axis was again parallel to the beam (D_{long}). Another important difference between the H and D atoms is that the **de**-coupling of electron and nucleus for the D case is higher. The decoupling can be quantified in terms of a critical field B_c which is 50.7 mT for H and 11.7 mT for D because of the smallness of the magnetic moment of the deuteron. The longitudinal target field of about 300 mT is 6× the B_c of H, but 26× the B_c of D, so the decoupling is about 4× better. One expects that in case of strong decoupling the target nuclei are weakly affected by random fields e.g. at the cell surface which tend to depolarize them.

As Paolo Lenisa, leader of the HERMES Target Group (1998–2005) stated in his target summary 2007: *In 2000 we had the ideal target, when the longitudinal deuterium target worked close to perfection. There was a continuous smooth running for 8 months when there was no temperature dependence between 90 and 40 K visible in the polarization of atoms and the degree of dissociation, meaning that there was no detectable recombination and wall depolarization.*

In the golden (D_{long}) Run in 2000 and before, the number of detected DIS events exceeded that of other runs considerably. The areal density was 2.1×10^{14} nucleons/cm^2 and the vector polarization $P_z = 0.845 \pm 0.028$. The error ΔP_z corresponds to a relative error of 3.3%, close to the 3% error envisaged in the TDR. Here we see the effect, as mentioned above, that in case of a "perfect" target cell, the polarization and fraction of molecules is nearly uniform, resulting in negligible sampling corrections and a low systematic uncertainty in the polarization measurement.

• **Operating the H/D target in a High Radiation Area:**
The H/D target had a complex vacuum system consisting of several
subsystems with many pumps, sensitive diagnostics, RF systems, gas
injection etc. (see Fig. 4.20), located together with the spectrometer
on a huge platform on rails. A two-storey electronic hut on the
same rails houses all the auxiliary equipment for the H/D target and
the HERMES detectors. In beam position, the platform sits behind
2000 tons of concrete shielding which guarantees a low radiation
level in the huge underground East Hall of HERA. For inserting the
HERA tram needed for work in the tunnel, the shielding could be
removed within a day. If there would have been an emergency with
HERMES preventing its operation as part of the HERA electron
ring, a possible scenario was to open the shielding, move the platform
back into the hall, and close the ring with a temporary beam pipe.
Fortunately, this situation never occurred in 13 years of running of
HERMES, thanks to the excellent support of the DESY technical
groups.

There were four experiments located around the HERA ring
which were served simultaneously by the circulating beams of protons
and electrons (H1 and ZEUS) or one beam only (HERMES with
electrons, HERA-B with protons; the other beams were by-passing
the experiments). Usually, the injected beams lasted for about 10 h,
after which the remaining beams were dumped and the machine
refilled. This kind of running took place typically for about 9 months
per year. One day per month the machine operation was paused
and the experiment could be accessed for routine service. At all
experiments, the components were controlled by remote computers
and their status displayed on various screens in the Control Rooms
where the shift crews supervised the data taking.

In case of a hardware failure, e.g. at the H/D target, the
coordinating experiment had to come to an agreement between the
experiments on how to proceed, either to wait for the next regular
access day, or to stop HERA operation and grant immediate access
for the relevant group which meant pausing the data taking of all four
experiments. There were several failures over the years of a critical
component, the dissociator of the ABS. Here, hydrogen gas was

injected into a water-cooled glass or quartz tube with a cold nozzle at the end, generating a beam of hydrogen atoms and molecules. A RF-discharge close to the nozzle was induced with a couple of hundred Watts of RF power, resulting in a high degree of dissociation close to 100%. Sometimes, thermal stresses caused the glass tube to develop a crack, the vacuum interlock acted and the ABS was shut down. The repair usually took 6–8 h, including the re-starting of the ABS.

A constant difficulty in operating the H/D target was that the sophisticated technology was known only within a small group of target experts and was not as wide-spread as detector and analysis expertise for particle-physics experiments are. The H/D target was made possible by a long R&D effort, as discussed in Chapter 3.6, beginning in 1985 with the FILTEX project. The first generation of students graduated in 1992–1995 (Hans-Günter Gaul, Kirsten Zapfe, Wolfgang Korsch and Friedemann Stock) during the phase when a working target had to be brought to DESY, ready for installation. New students were attracted and started working on problems associated with optimal running and target analysis with the completely new requirement of precise measurement of polarization by the target diagnostics (Bernd Braun, Christian Baumgarten, Mark Henoch). The students from the design phase had left the group, and there was a lack of experts on target hardware.

• **ABS Test Bench in the JADE Hall:** Therefore, in parallel to running the H/D target, a test stand was set up in the JADE hall of PETRA. The JADE experiment, which provided the name of this hall, is famous for the discovery of gluon jets. Here, hardware could be developed and tested before installation into the operational HERMES H/D target and, in addition, new students could be trained. This development was the subject of the thesis research of both Norbert Koch and Alexander Nass. The emphasis was on "cold" hydrogen atomic beams with high density for injection into the system of sextupole magnets. Two types of dissociators were studied: the RF-dissociator (discharge by RF of 13.56 MHz) which was applied up to 2000, and the Microwave-dissociator (discharge by

Microwaves of 2.45 GHz), a new development. Fortunately, in 1996 a new group (INFN Ferrara) joined and started to take over fabrication of the fragile target cells originally developed at Wisconsin. Using the Wisconsin ABS received from the group of Willy Haeberli, the Ferrara group was able to study the ABS technology, to train new students, and became a corner stone of the HERMES target group. In 1998, Paolo Lenisa (INFN Ferrara) took over leadership of the target group, following Geoff Court (Liverpool), who had made major contributions to the HERMES target design, together with James Stewart, in particular on the cryogenics and the superconducting longitudinal magnet.

> *As a veteran of the ABS development, Steffens remembers over the years several broken dissociator tubes which requested immediate repair and his presence as a hardware expert. After a five-hour train ride from Erlangen to Hamburg, he rushed to the East hall where the experiment was on hold. A short briefing with the grad student Phil Tait in charge of the ABS, then HERA operation was stopped and we had about half a day of access. When breaking the Interlock, everybody who entered the experimental area, took a key from the switch board at the entrance. After successful repair, the experimental area had to be carefully evacuated. All keys from the switchboard had to be put back before the interlock could be set again, enabling restarting the accelerator. Violation of these rules were treated at the highest (Director) level.*
>
> *The most delicate operation was to mount a new dissociator tube with aluminum nozzle fixed with Indium solder to the tube, and where several seals separating cooling water and vacuum had to be mounted properly. This happened in a noisy environment with a constant warning from the loudspeakers "The magnets are not connected to ground" and the feeling that four experiments with a thousand researchers were eagerly awaiting the end of our intervention. After leak chasing and pumping down, the vacuum system was started. When we were convinced that the system behaved normal we declared our OK, and the rest of the H/D target start-up was done during injection of a new beam. In the next hours, all parameters of the H/D target were carefully watched before the target went back to normal operation — all in all a very stressful operation, terminated by one or two beers in the DESY Cafeteria. As Steffens recollects, such operations were always successful.*

4.6.3 *Evolution of the HERMES spectrometer 1996–2000 and running*

While the HERMES experiment was under construction in 1994–1995, it was realized that the ability to detect hadrons in the final-state would allow new capabilities to study the origin of nucleon spin. In a paper of May 1991, Frank Close and Milner discussed the potential of semi-inclusive DIS which includes, besides the scattered electron, the detection of the leading hadron from the fragmentation process, reflecting the nature of the struck quark. There were preliminary discussions to implement a more sophisticated particle identification in the HERMES spectrometer but the paramount priority of installing the HERMES experiment in summer 1995 to begin data taking meant that such considerations were deferred. However, by the spring 1997 collaboration meeting, the HERMES spokesperson requested proposals for possible upgrades. These included

- Proposal for a dual radiator RICH by Argonne, Bari, Caltech, Rome and Gent.
- Recoil detector by NIHKEF, VU-Amsterdam, Erlangen and Yerevan.
- Measurement of the double spin asymmetry in charm leptoproduction by Frascati, Colorado, Heidelberg, Illinois, TRIUMF, Yerevan and Zeuthen.

4.6.3.1 *Dual RICH detector*

The physics case focused on semi-inclusive DIS, open charm production and the ability to detect the Λ particle. The impetus in time was provided by the ability of new *aerogel* radiator materials. These were hydrophobic, optically transparent and had a refractive index matched to the needs of the particles in the HERMES experiment. The founding groups in the RICH proposal were Argonne (built the threshold Cerenkov), Bari (a new group interested in HPD, advanced photon detectors), Caltech (interested in aerogel), Gent and Rome (interested in readout system). These were joined by DESY-Zeuthen (general infrastructure, cooling), Frascati (PMT experience and magnetic shielding), Tokyo (aerogel from Japan) and Yerevan (TOF).

The RICH proposal was based on the conversion of the existing threshold Cherenkov to a Ring Imaging CHerenkov detector. The Cherenkov effect occurs when a charged particle moving at speeds above the speed of light in the medium emits photons which are detected in Photo-Multiplier Tubes (PMT's). In threshold mode, one looks for the presence or absence of the light. The Cherenkov radiation is emitted in a cone centered on the particle's trajectory, and the opening angle is determined by the particle's speed. The rings are detected by a matrix of small PM's in hexagonal close-pack, 1934 per half. The intercept of cone and detector plane defines the rings which are detected and hence the particle's speed determined.

The RICH upgrade was realized on an impressively short timescale. From the initial discussion in December 1996, a proposal was approved by HERMES in March 1997, testing began in fall 1997, installation took place in May 1998 and the first rings were observed in August 1998. Figure 4.21 shows the rear of the PMT detector plane, and in Fig. 4.22 rings measured by the RICH are shown.

Fig. 4.21:   Photomultiplier tube (PMT) matrix of the RICH detector, upper half, seen from rear (*Source*: DESY).

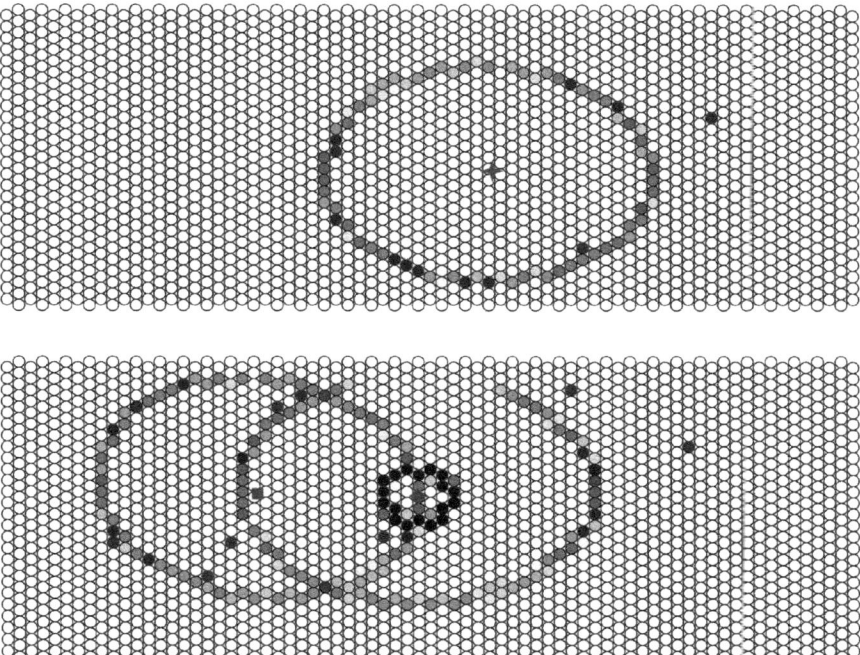

Fig. 4.22: Rings detected by the RICH (upper/lower detector); black dots: PMT's firing, grey: fitted rings. The smaller ring is due to the gaseous radiator (*Source*: HERMES Coll.).

The HERMES RICH was fundamental to the measurements involving hadron detection in the final-state, and thus essential for the groundbreaking measurements of semi-inclusive DIS and exclusive processes.

4.6.3.2 *Recoil detector*

The goal was to detect recoil protons from DIS in order to perform *exclusive* measurements, in contrast to *inclusive* (scattered electron, only) or *semi-inclusive* measurements (electron and leading hadron from fragmentation of the struck quark). A silicon test detector had been used located below the target cell with partial success. It helped for the design of the Λ-wheels (2001–2005), and later for a large-scale

(Si) detector with unpolarized target. The latter setup was used in the last phase of HERMES running in 2006–2007.

4.6.3.3 *Charm upgrade*

Besides a RICH to identify pions, kaons and protons, it comprised muon identification in a large acceptance, including tracks through magnet iron and an iron wall behind the calorimeter, a large-angle calorimeter in the front region and a downstream quadupole spectrometer to detect extreme-forward electrons. In the end, this upgrade did not yield a measurement of ΔG, the gluon contribution to the nucleon spin through charm production, but improved the detector and provided a higher level of understanding of the MC simulations fundamental for the analysis.

During the shut-down in 1997–1998, much of the detector upgrades were installed, including

- Iron wall for muon detection and shielding against background from the proton beam;
- Wide angle muon hodoscope;
- RICH detector.

In addition, the H/D target was converted to deuterium operation. This, together with the charm upgrade, gave access to deuteron and neutron spin structure functions and allowed for a powerful flavor decomposition. During the running 1998–2000, HERMES collected large data sets with vector- and tensor-polarized deuterium. A summary of operation of the polarized H/D target is shown in Table 4.1.

Large quantities of DIS and SIDIS data were obtained with nuclear targets (^{3}He, N_2, Ne, Kr, Xe). These targets were produced by injecting unpolarized gas directly into the target cell. They could have a considerably higher density as in the polarized case, where the density is limited by the ABS, and were applied in so-called end-of-fill runs to kill the remaining electron beam in a short time, thus resulting in high event numbers. In this way, several effects were discovered and studied, e.g. attenuation in the nuclear medium, the EMC effect and other nuclear effects. The understanding of the detector, radiative corrections, and the MC simulations had significantly improved. In summary, the period 1998–2000 turned out to be particularly successful.

4.7 HERMES Run II, 2001–2007

This period following the HERA luminosity upgrade, can be divided into two parts: (i) run IIa (2001–2005) with transverse polarized hydrogen target, and (ii) run IIb (2006–2007) with an unpolarized gas target and the recoil detector. The operation of the HERA collider was terminated on 30 June 2007.

4.7.1 *HERA machine upgrade and impact on HERMES*

By the late 1990s, the HERA electron–proton collider was in routine operation and the total luminosity acquired was approaching 100 pb^{-1} per year. Thus, a luminosity upgrade, known as HERA-II, was proposed in 1998 with the aim of achieving an integrated luminosity of 1000 pb^{-1}. Higher luminosity was attained by stronger focusing of the proton and lepton beams as well as better matching of the lepton beams. In addition, spin rotators at the two collider experiments, H1 and ZEUS, were added to allow those experiments to take data with polarized lepton beams, in addition to HERMES. As a consequence, the longitudinal magnetic fields at H1 and ZEUS were no longer perfectly compensated, reducing the attainable polarization.

The HERA-II upgrade implementation took place in 2001 and commissioning got underway later that year. However, unexpectedly large backgrounds in the ZEUS and H1 detectors were observed in 2002. A series of operational improvements were implemented. These included: additional collimators at the IP, better vacuum pumping in the vicinity of the IP, beam-based alignment of magnets, beam orbit feedback and automated beam optics corrections. By 2004, HERA was operating with about a factor of two increase in collider luminosity and more than 50% lepton beam polarization. By June 2007, when HERA operations ceased, the collider experiments had acquired an integrated luminosity of 500 fb^{-1}. However, from the point of view of HERMES, the HERA-II upgrade resulted in more than two years without data taking. The average HERA-II lepton polarization was lower, compared with HERA-I (1995–2000).

4.7.2 *Implementation and running with transverse hydrogen spin*

In run I, the helicity of the nucleon was measured with longitudinal polarization of the beam particles (electrons and positrons) and target particles (H, D, ^{3}He). A new feature pioneered by HERMES was the detection of leading hadrons which allowed for the determination of the flavor of the struck quark. This measurement is called Semi-Inclusive DIS (SIDIS). In run II, new observables, the so-called Single-Spin Asymmetries (SSA) have been studied with the H target polarized in transverse (vertical) direction, and by measuring the azimuthal angle ϕ of the hadrons to the spin direction around the beam axis and its harmonic content, like a $\sin(n\phi)$ dependence. Thanks to the enormous efforts of the theorists, many QCD effects could be related to the measured single spin asymmetries, making them the largest and fastest growing area of HERMES results. They will surely play a dominant role in future spin physics experiments at the EIC or the LHC.

Already during the previous running with longitudinal target spin, depolarization of the target by the RF field of the electron beam, known as *beam-induced depolarization,* had been studied systematically. This is demonstrated in Fig. 4.23 which shows the effect of the HERA beam on the rate in the target polarimeter as a function of the target magnetic field. In a strong field, i.e. in the right-hand side of the figure, the resonances are well separated, and a safe working point, without concern for beam-induced depolarization, could be found.

In the case of transverse spin, the situation is more difficult as a new class of so-called σ resonances with narrow spacing comes into play. To avoid them, one has to carefully choose the guide field to avoid those resonances. This requires a higher field quality than

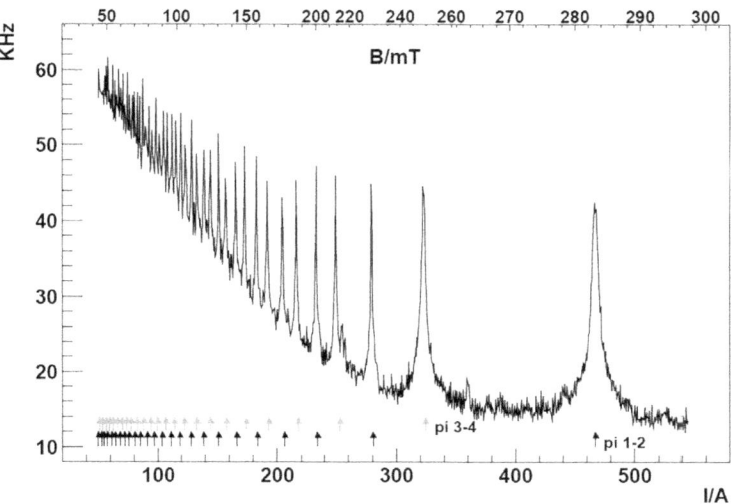

Fig. 4.23: Beam-induced (3–4) and (1–2) transitions with a longitudinally polarized hydrogen target as a function of the target field, proportional to the current I of the target magnet. The target would partly depolarize if the B-field would be kept at the positions of the peaks (*Source*: HERMES Coll.).

Fig. 4.24: Transverse magnet; chamber with target cell dismounted for field measurements. The lower pole with shims for improving the homogeneity is clearly visible (*Source*: DESY).

Table 4.1:   Summary of HERMES polarized target performance.

Run	Years	Target	DIS (Million)	Polarization
Run Ia	1995	$^3\text{He}_\parallel$	5	0.460 ± 0.02
Run Ib	1996–1997	$\text{H}_\parallel$	3.5	0.851 ± 0.033
Run Ic	1998–2000	$\text{D}_\parallel$	10.2	0.845 ± 0.028
Run IIa	2001–2005	$\text{H}_\perp$	~6	0.74 ± 0.06

that required for the longitudinal spin case. This is a challenge to achieve for a magnet with a large gap needed to accommodate the target chamber between the poles. A design was performed using an engineering program to optimize the magnetic field distribution. Special high-flux materials were used for the yoke and poles. The magnet was designed by Tom Wise of the Madison group and built at the Wisconsin University workshop. It was then empirically optimized by adding additional shims to the poles, as shown in Fig. 4.24. With the help of correction coils (not shown), the required field quality could be achieved.

Table 4.1 lists the performance of the ^3He target and H/D target with both longitudinal and transverse spin orientations.

Chapter 5

HERMES Scientific Output

5.1 Overview

Note: This chapter contains a significant number of technical terms. The reader of non-technical background may wish to consult the definition of technical terms in Appendix A.

In Chapter 4, we saw that there were two distinct phases to the operation of the HERA collider: the initial phase HERA-I from 1992 to 2000, and the second phase HERA-II from 2002 to 2007 after the collider luminosity upgrade and installation of spin rotators for the H1 and ZEUS experiments. The HERMES data taking took place in three distinct running periods from July 1995 to June 2007:

- *Run I: 1995–2000* — Here, the focus was on spin-dependent electron and positron scattering from longitudinally polarized targets of helium-3, hydrogen and deuterium. These data yielded high-precision results on inclusive and semi-inclusive polarized DIS. They have confirmed and extended our understanding of the contribution of quark spins to the spin of the nucleon.
- *Run IIa: 2002–2005* — Here, the focus was on spin-dependent electron and positron scattering from a transversely polarized hydrogen target. The scientific thrust was to determine the polarization of the quarks in a transversely polarized nucleon and to explore the transverse momentum distributions of the quarks for the first time.

- *Run IIb: 2006–2007* — Here, the focus was on spin-dependent electron scattering using a recoil detector to determine generalized parton distributions and the role of orbital angular momentum in the origin of the proton spin.

Further, from 1995 onward, data were also collected with various other unpolarized gas targets (hydrogen, deuterium, both helium isotopes, nitrogen, neon, krypton, and xenon) to study how the struck quark transforms into the detected hadron (known as the *hadronization* process).

In this chapter, we describe how HERMES made a pioneering set of measurements to unravel the quark and gluon polarizations in the proton. Key to this was the ability to detect mesons in coincidence with the scattered electron in the violent DIS process where the proton is smashed into pieces. This was made possible by the RICH detector described in Chapter 4. As HERMES carried out these measurements in the late 1990s and early 2000s, theoretical physicists developed an entirely new perspective on understanding proton structure. Transversely polarized targets were essential and the HERMES data taking in 2002–2005 yielded high quality data which, as we shall see, were key to making progress. Further, an entire new class of measurements was identified where the proton was left intact in the scattering process and, further, a new particle (photon or meson) was produced in addition to the scattered electron. HERMES pioneered measurement of these *exclusive* processes, in particular where a photon was produced. Finally, the HERMES data on nuclei proved to be important in understanding relativistic heavy-ion collisions aimed at investigating hot, dense QCD matter.

5.2 Target Spin Along Beam Direction

5.2.1 *Detecting only scattered electron*

We have seen that HERMES was motivated by the desire to understand how the intrinsic angular momentum of the proton arises from its fundamental constituents, the quarks and gluons. HERMES used electrons to scatter from charged quarks in the proton. In the mid-1990s, when HERMES started data taking, the principal focus was on measurements where only the scattered electron was detected, known as *inclusive* scattering. Such measurements could be interpreted using sum-rules over the measured structure functions. In response to the observed violation of the Ellis-Jaffe Sum Rule on the proton by the EMC experiment, new measurements were underway at CERN by the SMC experiment and at SLAC in End Station A. The measurement of the neutron spin-dependent structure function and its integral were a principal goal in 1995.

By May 1994, HERMES had installed a test experiment which consisted of: the calorimeter setup: target chambers with wakefield suppressors and one element of each detector component. Studies with single and colliding beams showed that the collimators removed the synchrotron radiation and the proton-induced background. The electron beam lifetime was not influenced by gas introduced into the target cell. The electron beam backgrounds seen were subsequently understood and minimized. This test experiment was crucial to the success of the 1995 installation of the complete experiment.

In summer 1995, when the HERMES collaboration began data taking with a polarized ^{3}He target, the primary objective was to acquire high-quality data on the the neutron so that the Ellis-Jaffe Sum Rule for the neutron and the Bjorken Sum Rule could be checked experimentally. Both experiment E-154, in End Station A at SLAC, and HERMES at DESY were taking data with polarized ^{3}He targets that summer. Since the demise of the PEGASYS proposal at SLAC, and motivated by the surprising results from EMC, SLAC had developed renewed interest in spin-dependent DIS in End Station A. Emlyn Hughes, son of Vernon Hughes and a physics faculty member at Caltech, was a leader of this effort. At this time, Vernon's focus on spin-dependent DIS was at CERN with the Spin Muon Collaboration. Thus, there was a father-son competition to understand the spin structure of the nucleon.

At DESY in the hot summer of 1995, the East Hall of HERA was a hive of activity with the commissioning of the HERMES experiment in full swing. Milner recalls walking into the HERMES control room in June and coming face to face with the world wide web for the first time. HERMES had a website. Also, the Microsoft Powerpoint program for presentations appeared. Michael Düren was an early user of this which still required plastic transparencies shown using an overhead projector. The arXiv, which had been in existence for a few years, really became the universal tool in 1995 with the arrival of the internet. The internet allowed the downloading of large postscript files from the arXiv. These developments truly revolutionized access to the latest scientific papers.

The introduction of the HERMES experiment to the HERA circular accelerator initiated several months of careful tuning of the stored electron beam, which had been optimized for the existing ZEUS and H1 collider experiments. A carefully designed set of beam collimators upstream of the HERMES experiment shielded the sensitive detectors from an intense flux of synchrotron radiation generated in the magnets. This included both a fixed collimator and a movable collimator. When the electron beam was injected in HERA, the detectors were made less sensitive, the collimators were opened, the beam was injected and a stable orbit was established. Once conditions were optimized, the movable collimator was closed, the detectors were brought into regular operation and data taking commenced. Data taking continued for several hours while the beam intensity decayed. At a low circulating current, the beam was dumped and the cycle was repeated.

In the initial year of running, HERMES had chosen to install the polarized ^{3}He gas target. This was motivated by the desire to measure the spin asymmetry on the neutron for the first time. As Emlyn Hughes and his colleagues were carrying out a similar measurement at SLAC there was a race to determine the unmeasured inclusive neutron asymmetry.

Commissioning and optimization of the HERMES detector performance was a major activity from late spring through summer 1995. In particular, tuning of the stored HERA electron beam to minimize background in the tight collimation system and HERMES target region took many weeks to achieve an adequate performance so that data acquisition could proceed routinely.

Happily, by August 1995, HERMES was starting to take spin-dependent electron scattering data from polarized ^{3}He and this continued until late November. HERMES accumulated 2.7 million inclusive DIS events from polarized ^{3}He in the initial 1995 run, and verified that the sum rule was consistent with expectations within $\pm 35\%$. By comparison, E154 collected 100 million DIS events and determined the neutron sum rule to $\pm 20\%$. Undoubtedly, E154 won the competition of 1995 to determine the neutron asymmetry from ^{3}He but the results were entirely consistent with expectations and, most importantly for HERMES, a new type of experiment was operating successfully.

By the end of 1995, Richard Milner was Spokesman of HERMES and the collaboration was faced with a decision on the choice of target for 1996 running. While it would have been more straightforward to continue data taking with the ^{3}He target, scientific considerations made it clear that installing the ABS target was the correct choice. Thus, the winter 1995–1996 shutdown was hectic with removal of the ^{3}He target and the installation of the ABS/hydrogen target for the first time (January 1996). In addition, for Steffens the transition from Heidelberg to Erlangen university and new teaching and administrative responsibilities were on the agenda, including the move of his family to Erlangen with three school kids. In October, he stepped down as Technical coordinator, handing over to Armand Simon. Geoff Court (target coordinator) and Jim Stewart from U. Liverpool played a leading role in the installation of the polarized hydrogen target for 1996 running, supported by numerous members of the German H/D target groups Heidelberg, Marburg and Munich, in particular Friedemann Stock, Peter Schiemenz and Kirsten Zapfe, and by the local DESY technical groups.

As summarized in Table 4.1, in 1997 and 1998, HERMES took data with a longitudinally polarized hydrogen target and in 1999 and 2000 on a longitudinally polarized deuterium target. The data were of high quality and Fig. 5.1 shows the HERMES determination of the spin-dependent structure functions of the proton $(g_1^p(x))$ and deuteron $(g_1^d(x))$ as a function of x. The functions $g_1(x)$ are

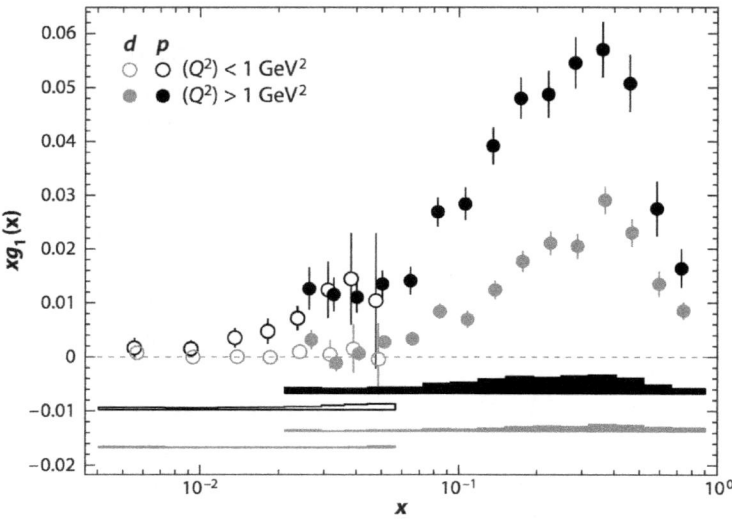

Fig. 5.1:   Measurements of the spin-dependent structure functions of the proton (p) and deuteron (d) as a function of x by HERMES from [10]. The functions $g_1(x)$ are derived directly from the longitudinal spin asymmetry $A(x)$, defined together with x in Section 1.5 (*Source*: HERMES Coll.).

derived directly from the longitudinal spin asymmetry $A(x)$, defined in Section 1.5. The deuteron contains an additional neutron whose structure function is opposite in sign to that of the proton. This is why at low $x < 0.05$, the deuteron structure function is close to zero.

With the data of Fig. 5.1, the contribution of the quarks to the spin of the proton can be determined and HERMES determined this to be about one-third within an uncertainty of about 10%. The rest must be attributed to gluons and orbital angular momenta.

5.2.2   *Detecting the struck quark*

In March 1997, *Science* magazine featured the new HERMES experiment in an article in its Research News section. MIT theorist Bob Jaffe, and HERMES collaboration leaders Klaus Rith and Richard Milner were quoted. HERMES had verified that quarks account for about 25–30% of the proton's spin. Looking to the future, the article reported that HERMES was about to build a new detector that would allow the identification of the hadrons resulting from the struck quarks. Interestingly, the article caught the eye of Burton Richter at Stanford who wrote a letter to *Science* some weeks later. A main point of Richter's letter was "The precision of the most recent SLAC data is far better than that achieved at HERMES." However, he went on "The real power of HERMES lies in its potential to capture the debris of the protons and neutrons after they are struck by high energy positrons from the HERA ring."

While the HERMES experiment was being approved and was under construction, the collaboration had realized the significant physics potential in detecting hadrons in coincidence with the scattered lepton, so-called *semi-inclusive deep inelastic scattering* (SIDIS). As shown schematically in Fig. 5.2, in DIS the electron (or positron) strikes a quark which, in the simplest approximation, materializes as a detectable quark–anti-quark pair (called a *meson*). Detection of the meson in coincidence with the scattered lepton, can allow isolation of events that, with high probability, contain the struck quark. This can allow tagging of the flavor of the quark that was involved in the DIS process. Consequently, the contribution

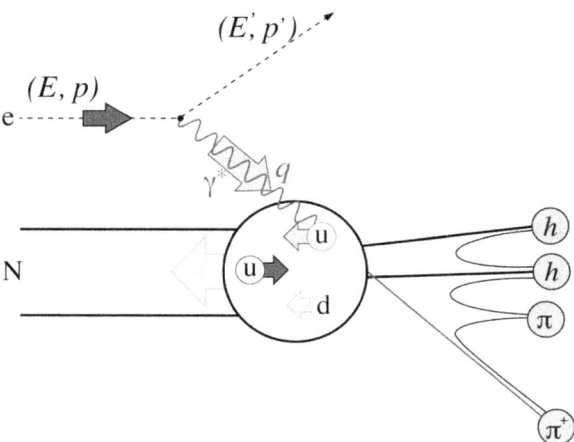

Fig. 5.2: Schematic diagram for semi-inclusive deep inelastic electron scattering from the proton (*Source*: Marc Beckmann).

of the cross-section or spin asymmetry due to scattering from the quarks of a particular flavor can be isolated, in this approximation. Further, the process by which the struck quark materializes as a meson is one of the most fundamental in experimental study of the strong interaction at high energy. It is known as *hadronization* and it connects the experimental world of detectable particles with the hidden world of quarks and gluons described by Quantum Chromodynamics. The simplest description of the hadronization process introduces an extension of the quark parton model where new *fragmentation functions* enter. While some data had been taken which agreed with this model, there were many open questions for HERMES to pursue, particularly those pertaining to spin.

These scientific considerations had been the primary motivation for the PEGASYS collaboration at SLAC to consider an internal target experiment in the PEP ring and the EMC experiment at CERN had carried out a program of unpolarized measurements where the scattered muons were detected in coincidence with hadrons. However, HERMES had the possibility to measure spin asymmetries in such experiments for the first time. Thus, as the HERMES experiment started data taking in 1995, upgrades of the detector which would allow clean particle identification were underway.

The central element of the upgrade was a Ring Imaging CHerenkov (RICH) detector. This was essential to hadron detection and its realization is described in Chapter 4. Harold (Hal) Jackson from Argonne National Laboratory, was a prime mover in developing both the science case and the technical realization of the upgrade.

A significant effort went into developing the formalism for extracting the polarizations of the quarks in terms of valence vs. sea and as a function of flavor. The MIT group led by Richard Milner and Robert Redwine played a central role in this development with work by Michael Niczyporuk (senior thesis 1997) developing the concept of quark flavor purity, post-doctoral researchers Eppo Bruins and Arthur Mateos, culminating in the 1999 Ph.D. thesis work of Bryan Tipton on extracting the first polarized parton distributions in the proton from HERMES data in 1995, 1996 and 1997. Using both inclusive and semi-inclusive HERMES data, the

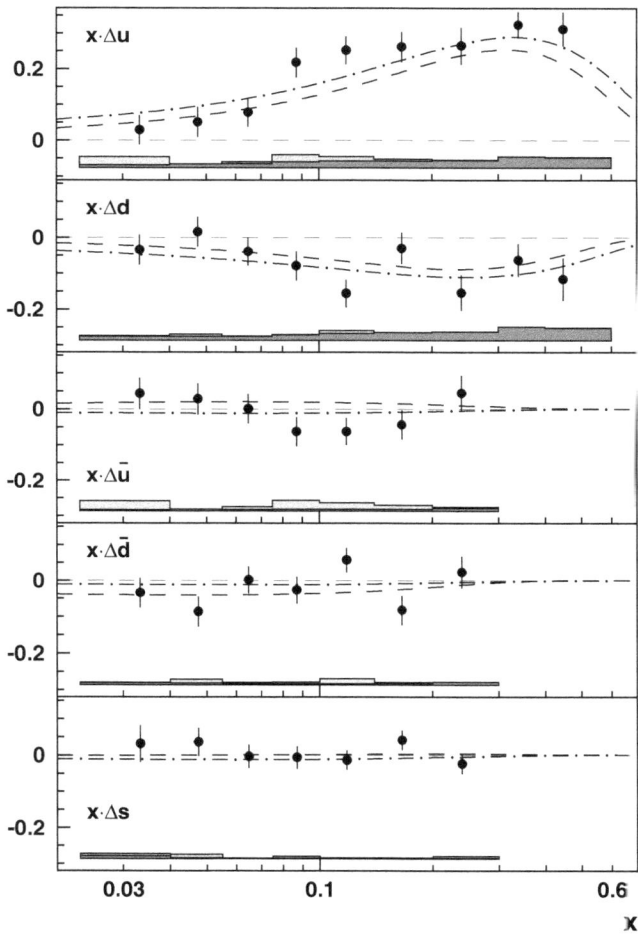

Fig. 5.3: The quark helicity distributions for *up, down, anti-up, anti-down* and *strange* quarks as a function of x as determined from the HERMES DIS and SIDIS data.[11] x is defined in Section 1.5. The curves represent the theoretical expectations (*Source*: HERMES Coll.).

up quarks were found to be polarized in the same direction as the proton but the *down* quarks were polarized in the *opposite* direction. Figure 5.3 shows the resulting quark polarizations based on longitudinally polarized hydrogen data in 1997 and 1998 and longitudinally polarized deuterium data in 1999 and 2000.

Experimental information about the contribution of *strange* quarks ($s(x)$) and antiquarks ($\bar{s}(x)$) to the structure of the proton as a function of x is surprisingly scarce. Most of the experimental constraints are based on measurements of oppositely charged muon pairs in deep-inelastic neutrino and antineutrino scattering. HERMES performed the first extraction of $S(x) \equiv s(x) + \bar{s}(x)$ from the multiplicity of charged kaons in semi-inclusive DIS from a deuteron target. The measured kaon multiplicity cannot be reproduced using the standard parametrization for strange quark distribution functions. The HERMES data for $xS(x)$ show sizable and increasing strength below $x = 0.1$. Above $x = 0.1$, the HERMES data for $xS(x)$ are consistent with zero.

Finally, HERMES made a determination of the gluon polarization using spin-dependent inclusive hadron production at high transverse momentum from a deuterium target. At an average x value of 0.22, the gluon polarization was found to be small and consistent with zero.

5.3   Detecting Photons

In the late 1990s, HERMES embarked on the measurement of high energy photons in coincidence with the scattered lepton. The motivation was to access for the first time Generalized Parton Distributions (GPDs). The process by which the scattered electron left the proton intact and a photon was additionally emitted, so-called Deeply Virtual Compton Scattring (DVCS), was identified as particularly promising measurement by which to access the new GPDs. Theorists had identified DVCS as a new, exciting way to probe the structure of the proton.

HERMES immediately sought to extract information on this new process. The Armenian physicist Moscow Amarian was a key driver of this effort. In the initial days of HERMES, Moscow had worked as a member of the Armenian group from Yerevan. Subsequently, he was employed by NIKHEF and by 1997 he was in Rome working in Salvatore Frullani's group. Moscow was an inventive physicist who tirelessly sought out new observables in the HERMES data. He had observed open charm production and worked on a determination of gluon polarization using dihadron production with Jeffrey Martin from MIT. While in Rome, Moscow realized that the interference between the standard radiation (so-called Bethe-Heitler) and the hard photon process (so-called Deeply Virtual Compton Scattering) should be observable in HERMES. In particular, the beam spin asymmetry should be particularly sensitive. Frullani encouraged Amarian's DVCS analysis in spite of skepticism within the HERMES collaboration. While in Rome, Amarian came to take shifts on the CLAS experiment at Jefferson Laboratory, Newport News, VA. When he visited, he related his work on DVCS at HERMES and the CLAS collaboration pursued measurement of DVCS at 4.25 GeV. In August 2000, Harut Avakian gave a talk at the Gordon Conference in New Hampshire, USA on the HERMES spin asymmetries for DVCS. The Gordon conference allows participants to present preliminary results without fear of them being cited. Many physicists from the CLAS collaboration were in attendance. Moscow Amarian gave the first public report of a measurement of DVCS by HERMES at 27.5 GeV at the workshop on Skewed Parton Distributions at DESY in September 2000. COMPASS, H1 and ZEUS also reported DVCS observations there and CLAS was working on it. In the next month, Amarian presented the HERMES results at the 14th International Spin Physics Symposium in Osaka, Japan in October, 2000. Harut Avakian took up a position at Jefferson Lab in 2001 and immediately went to work on the CLAS analysis of DVCS data. The papers reporting the HERMES and CLAS observations of the DVCS process were published back-to-back in the same 29 October 2001 issue of Physical Review Letters. Figure 5.4 shows the HERMES DVCS data.

5.4 Target Spin Transverse to Beam Direction

HERMES began data taking at a time when theorists were beginning to become interested in the transverse structure of the proton. Piet Mulders and Daniel Boer in Amsterdam, Aram Kotzinian in Yerevan, Mauro Anselmino in Torino, Robert Jaffe and Xiangdong Ji at MIT were each working on theoretical aspects that built on earlier work

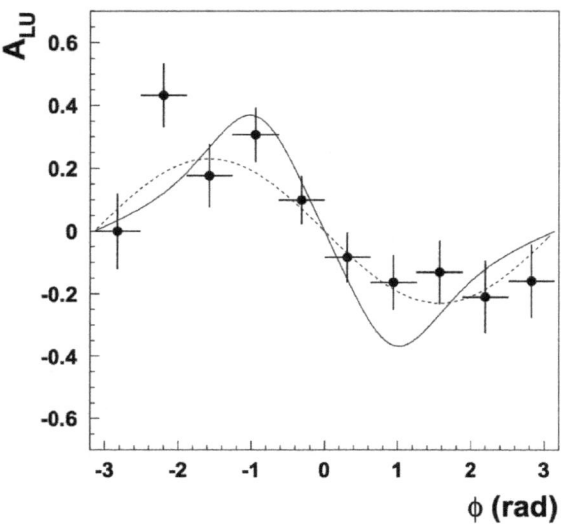

Fig. 5.4: First measurement by HERMES of the beam spin asymmetry for hard electroproduction of photons as a function of the azimuthal angle[12] (*Source*: HERMES Coll.).

by John Collins and Dennis Sivers and others. Harut Avakian recalls working on the analysis of HERMES data on single-spin asymmetries associated with the azimuthal distribution of hadrons but meeting with skepticism from the HERMES collaboration leadership. It was only in 1999 that he received permission to present results for the first time at the DIS99 conference held at DESY-Zeuthen. He recalls that Bob Jaffe was excited about the possibility to access information on transversity for the first time. This became the start of a major push by HERMES to pursue this physics.

In the years 2002–2005, the HERMES experiment took data with polarized proton targets polarized transverse to the incident beam direction. The RICH detector provided the ability to cleanly detect hadrons in coincidence with the scattered electron. These measurements allowed access to previously unmeasured aspects of the proton structure for the first time. A key aspect was the measurement of the scattering asymmetry as a function of the azimuthal angle. Previously, with longitudinal target spin, the single-spin asymmetry was measured by HERMES as a function of the azimuthal angle for the first time. Such asymmetries had previously been observed in polarized proton scattering at Fermilab and were later seen in the STAR experiment at RHIC. They must arise from a new spin–orbit interaction either in the fragmentation process or in the proton itself. The spin-orbit interaction in the fragmentation is known as the *Collins effect*. Interaction between the nucleon spin and the quark orbital motion is termed the *Sivers effect*.

A complete description of the proton's structure requires three quark distribution functions that survive integration over intrinsic transverse momenta. These are the unpolarized quark distribution function, the quark helicity distribution function, and the novel *transversity* distribution function. There are five additional transverse momentum-dependent distribution functions that do not survive the integration. They were essentially unexplored experimentally, and HERMES performed pioneering measurements to access all of them. One example is the time-reversal-odd *Sivers* distribution function, which describes the distribution of unpolarized quarks in a transversely polarized nucleon and arises from a correlation between the momentum direction of the struck quark and the spin direction of its parent nucleon. It is of special interest as it can be related to the orbital angular momenta of quarks. One way to study this distribution function is via the azimuthal angular asymmetry in the distribution of hadrons produced in semi-inclusive DIS from a transversely polarized target. It gives rise to a modulation in the azimuthal distribution.

HERMES carried out measurements of pion electroproduction with a transversely polarized target. By determining the azimuthal asymmetry as a function of kinematic variables, HERMES obtained the first evidence for non-zero Collins function and transversity. The effects were positive for positively charged pion and opposite in sign and larger for the negatively charged pion. Further, HERMES obtained the first evidence for the non-zero Sivers function, as seen in Fig. 5.5. This is direct evidence of a non-zero quark orbital angular momentum inside the proton. However, at present it is not possible to quantitatively relate the magnitude of this asymmetry to the fraction

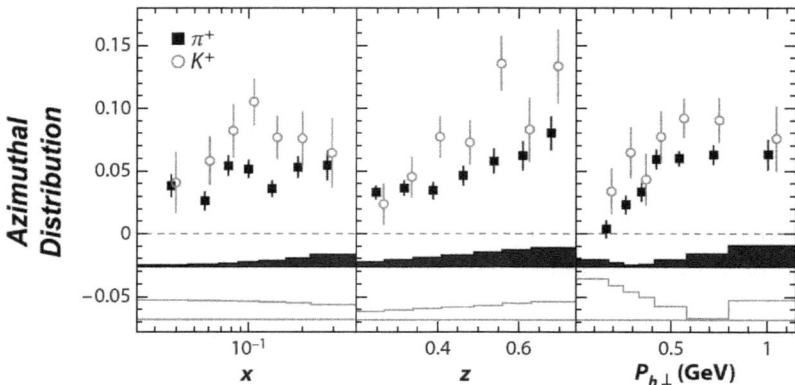

Fig. 5.5:   Results for the azimuthal distribution (known as the Sivers moment) for positive pions and kaons obtained with a transversely polarized hydrogen target by HERMES from Ref. [10]. The quantity x is defined in Section 1.5; z is the energy of the pion or kaon after the hard scattering; and $p_{h\perp}$ is the momentum of the pion or kaon transverse to the beam direction (*Source*: HERMES Coll.).

of nucleon spin that can be attributed to orbital angular momenta of quarks. In addition, HERMES measured a significant Sivers effect for charged kaons.

The HERMES results on transverse spin generated great interest worldwide in the QCD community. They pioneered the use of leptoproduction to study the fundamental structure of the proton in a new way. Theorists were excited to see data that could be used to constrain the modeling of theses new phenomena. Further, the HERMES collaboration, with Naomi Makins playing a leading role, made effective animations of these new observables which were widely used in the community.

5.5   What about Nuclei?

From the earliest running with the test experiment in 1994, the HERMES target offered the possibility to flow unpolarized gases. Further, it was realized that at the end of a HERA electron beam fill, when the collider detectors ZEUS and H1 had turned off, a considerable quantity of data could be acquired in a relatively short running time

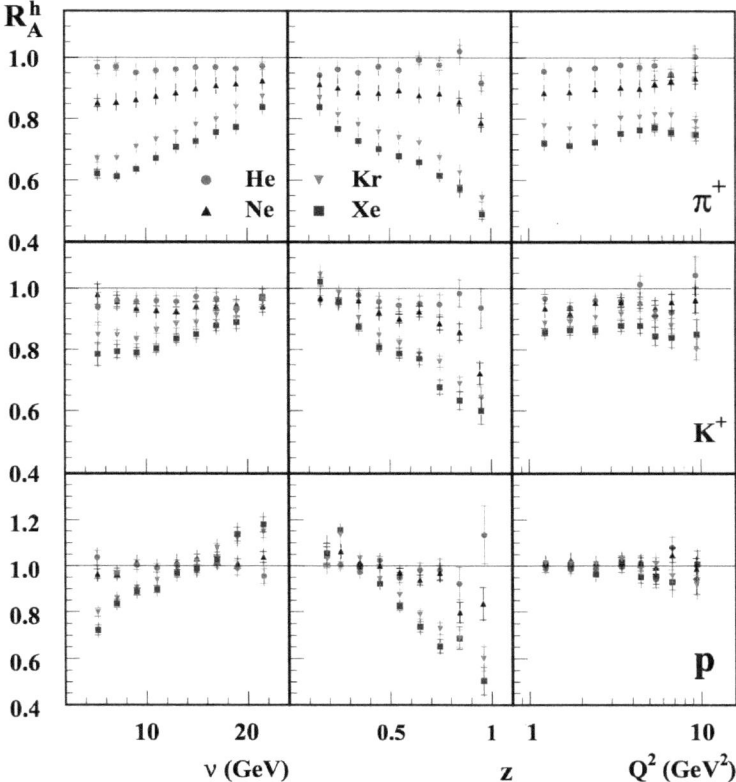

Fig. 5.6:   HERMES data on the kinematic dependence of production of π^+, K^+ and protons from helium, neon, krypton and xenon targets.[13] ν is the energy lost by the incident electron; z is the energy of the pion, kaon or proton after the hard scattering; and Q^2 is defined in Section 1.5 (*Source*: HERMES Coll.).

by increasing the unpolarized gas density. Thus, HERMES could take a large amount of both inclusive and semi-inclusive DIS data from nuclei (Fig. 5.6). This had been the primary motivation of the PEGASYS proposal at SLAC, described in Chapter 3.

The HERMES data on SIDIS from nuclei (Fig. 5.6) provided important information on quark energy loss as it traversed the nucleus. This energy loss measurement provided a baseline to which the quark energy loss in the hot, dense matter formed in heavy ion collisions could be compared. The Relativistic Heavy Ion Collider at

Brookhaven National Laboratory, Upton NY began taking data in 2000 and observed large energy loss by struck quarks in the hot, dense matter formed in these high-energy collisions. Using the HERMES data as a comparison, theorists were able to determine that the energy loss in the hot, dense matter was about 30 times that in ordinary cold nuclei. This provided strong evidence that this new matter had properties very different from what had been observed before.

As the DIS data on nuclei were analyzed, an unexpected suppression of the nuclear inclusive cross section at low x was observed by HERMES. It seemed inconsistent with previous measurements at CERN but there was the possibility of a kinematic dependence due to the lower energy of HERMES. The effect withstood persistent scrutiny and investigation by the collaboration and it started to be discussed as the *HERMES effect*. Some theorists wrote a paper in 2000 explaining it! However, there were hints that radiation of a high-energy photon by the beam could be problematic. After a year of study by the HERMES collaboration, with the Canadian physicist Andy Miller playing a leading role, it was concluded that such a high-energy photon could shower in the beampipe and produce a large number of radiative events, which could mimic real DIS events. The radiative correction codes over-corrected for this and this was the origin of the *HERMES effect*. Once this class of radiative events were correctly taken into account, the HERMES nuclear DIS data were in good agreement with the world's data.

5.6   Role of Theory

As we explained in the early chapters of the book, the Standard Model theory of the strong force, namely Quantum Chromodynamics, had been developed already in the 1970s. However, it was not possible to carry out accurate QCD calculations of the reactions being measured by the HERMES experiment. Starting in the late 1980s, a new generation of young theoretical physicists opened up novel avenues that, together with HERMES and other experiments, resulted in a new twenty first century view of the proton. There

Fig. 5.7: Chinese-American physicist Xiangdong Ji from the University of Maryland and Jiao Tong University, Shanghai.

were two theory post-docs at Caltech at the time that the HERMES experiment was being invented; Xiangdong Ji and Andreas Schäfer, who both went on to have a large influence on the HERMES experiment.

Xiangdong Ji, pictured in Fig. 5.7 is a brilliant Chinese-American physicist, who, first at MIT and later at the University of Maryland, has had a large impact on how we understand and study the proton. He has led the way in viewing the proton as a three-dimensional object that can be imaged using electron scattering. To this end, he developed a whole new, more general classification of the distribution of quarks and gluons. In 1996, his paper first pointed out the importance of the DVCS reaction, discussed in Section 5.3 above. He worked out a method to understand the origin of the proton spin. Further, he developed a novel scheme to calculate the quark and gluon distributions using large-scale computer simulations.

Andreas Schäfer returned to Europe, took a faculty position at the University of Regensburg and joined the HERMES collaboration. He worked with the HERMES experiment to produce the important scientific results described here. Klaus Rith and Andreas Schäfer wrote the article *The Mystery of Nucleon Spin* in the July 1999 edition of the magazine *Scientific American.*[14] In Amsterdam,

Piet Mulders and his students Daniel Boer and Alessandro Bacchetta worked on a general formalism to understand semi-inclusive DIS processes with spin. Aram Kotzinian in Yerevan also worked on polarized DIS. Matthias Burkardt and Markus Diehl worked on understanding the new, more general parton distributions. Mauro Anselmino in Torino and his Italian colleagues worked on single spin asymmetries and Marco Radici in Pavia worked on transverse spin distribution in the proton. Werner Vogelsang and his colleagues developed powerful global analyses of world data to understand what could be learned with new experiments. In Imperial College London, Elliot Leader worked on resolution of the spin crisis and wrote the classic textbook *Spin in Particle Physics*. Bob Jaffe, the co-developer with John Ellis of the sum rules that bear their name as discussed in Chapter 2, worked on the transverse spin structure, and a general framework in which the contributions of the quarks, the gluons and orbital angular momentum combine to produce the proton spin.

Another important development in the last several decades has been the increasing power and relevance of simulations of QCD using the world's most powerful computers. Here space and time are discretized on a lattice and so-called *lattice QCD* calculations of the quark and gluon structure of the proton and nuclei are carried out. Most of these calculations have been carried out in the rest frame of the proton but Xiangdong Ji has proposed an ingenious scheme to perform such calculations in the reference frame in which high-energy electron scattering data are interpreted in terms of QCD.

In the years 1990–2007, the HERMES experimentalists and QCD theorists had a productive, symbiotic relationship which quickly spread to the other experiments at SLAC, CERN, BNL and Jefferson Lab. The language used by physicists to describe the proton structure changed greatly between the pre-HERMES years and post-HERMES years. This can be seen directly by looking at conference proceedings and review articles.

5.7   Scientific Impact of the HERMES Experiment

The HERMES collaboration published 82 papers in scientific journals between 1997 and 2019. In this period, more than 135 Ph.D. theses were written on HERMES science. As we will discuss in Chapter 7, the impact of HERMES-trained physicists on academia, research and industry around the world has been considerable. Among the most cited of the HERMES papers are the measurement of single-spin asymmetries on a transversely polarized hydrogen target and the measurement of inclusive spin-dependent structure functions on the proton and neutron.

Firstly, HERMES was technically innovative and demonstrated the effectiveness of realizing experimentally the ideal lepton-nucleon scattering process with polarization of both lepton and nucleon. This is powerful, particularly for final-states where hadrons are detected. Previous fixed-target experiments used polarized targets where large amounts of extraneous unwanted materials are present, necessitating large corrections. This approach to the ideal measurement is now established in the future electron-ion collider with significantly higher collision rate, at much higher energies and over a much larger kinematic range.

HERMES accomplished its stated goal of determining the flavor-dependence of quark polarizations for the *up*, *down* and *sea* quarks. Together with other experiments at SLAC, CERN and JLab, we have established with high confidence that the quarks contribute about 30% of the proton's spin. We have further experimental data that support a significant contribution of the gluons and the role of orbital angular momentum, but these conclusions are not as definitive as those established using high-energy lepton scattering.

Most significantly, in the years 1990–2010, HERMES was a central player in a revolution in how physicists have come to think about the structure of the proton and nuclei in terms of the fundamental quarks and gluons of QCD. This was driven by theorists and involved other experiments, but HERMES made pioneering measurements involving hadron detection and transverse spin which have been essential to our twenty-first century view of the proton as an extremely relativistic, three-dimensional, bound system of quarks and gluons which can be described by a five-dimensional quantum mechanical function, the so-called *Wigner function*. By careful measurement of very specific observables that involve both violent destruction of the proton as well as non-violent scattering where the proton is left intact, snapshots of the proton in the plane transverse to the beam direction can be obtained which give us information on aspects of the Wigner function.

In conclusion, HERMES opened a powerful, new experimental avenue to the study of the fundamental structure of matter. The international HERMES collaboration built on the considerable strengths of its diverse institutions to utilize the unique HERA accelerator at DESY to carry out definitive measurements of the origin of the nucleon spin. HERMES played a major part in the development of the twenty-first century picture of the proton.

Chapter 6

Building on the Legacy of HERMES

6.1 Introduction

The HERMES experiment was a significant part of a transformative evolution, involving both experimentalists and theorists, that changed how we view and discuss the fundamental structure of the proton. The HERA collider experiments H1 and ZEUS in studying proton structure in a different way to HERMES also had an enormous impact, particularly with respect to understanding the role of gluons. Concurrent and subsequent experiments have carried forward frontier research into the fundamental structure of matter over the last two decades. Here, we describe how our current understanding of the fundamental structure of matter has built on HERMES and led to an exciting new future path well into the twenty-first century.

6.2 SLAC End Station A Experiments

We have seen that HERMES had origins at SLAC, when the effort to develop an internal target program at PEP was not pursued. However, subsequently an important and successful series of experiments was carried out in End Station A there using the polarized electron beam from the two-mile linac. These experiments employed dynamically nuclear polarized targets for the proton and deuteron (experiments E143 and E153) and drove the development of a high density polarized ^{3}He target using spin exchange optical pumping (experiments E142 and E154). Taking advantage of the energy doubling developed for the SLAC Linear Collider, these experiments were able to take data at 48.3 GeV and beam polarizations of about 80%. The experiments provided precise determination of the inclusive structure functions for the proton and neutron. The precision and kinematic reach in inclusive scattering achieved at SLAC could not be matched by HERMES and this further emphasized that the focus of HERMES should be on semi-inclusive measurements, where the hadron is detected in coincidence with the scattered electron. The collaborations and the polarized electron and target technologies were subsequently used at Jefferson Lab to great effect.

6.3 HERA Collider Experiments H1 and ZEUS

While the HERA electron-proton collider was constructed mainly to look for evidence of quark and lepton substructure, it's legacy has had an enormous impact on our understanding of the quark and gluon structure of the proton. HERA's ability to access new values of low x and high Q^2 allowed the precise determination of the quark distribution and, for the first time the gluon momentum distribution, over an unprecedented range in spatial resolution and shutter exposure time, as discussed in Section 1.5. Figure 6.1 shows an overview of the HERA data. The success of the HERA collider in gaining new insight into the fundamental structure of matter has been a significant motivation to the QCD community worldwide in determining the next-generation accelerator. For example, several key architects of the planned Electron-Ion Collider in the US had previously worked at DESY.

6.4 Jefferson Laboratory

The Continuous Electron Beam Accelerator Facility (CEBAF) (later to be named Thomas Jefferson National Accelerator facility) in Newport News, VA is the flagship facility in the US to understand the quark and gluon structure of matter using electron scattering. It was initially conceived in the 1970s and construction of a 4-GeV design commenced in 1987. The initial design was based around a

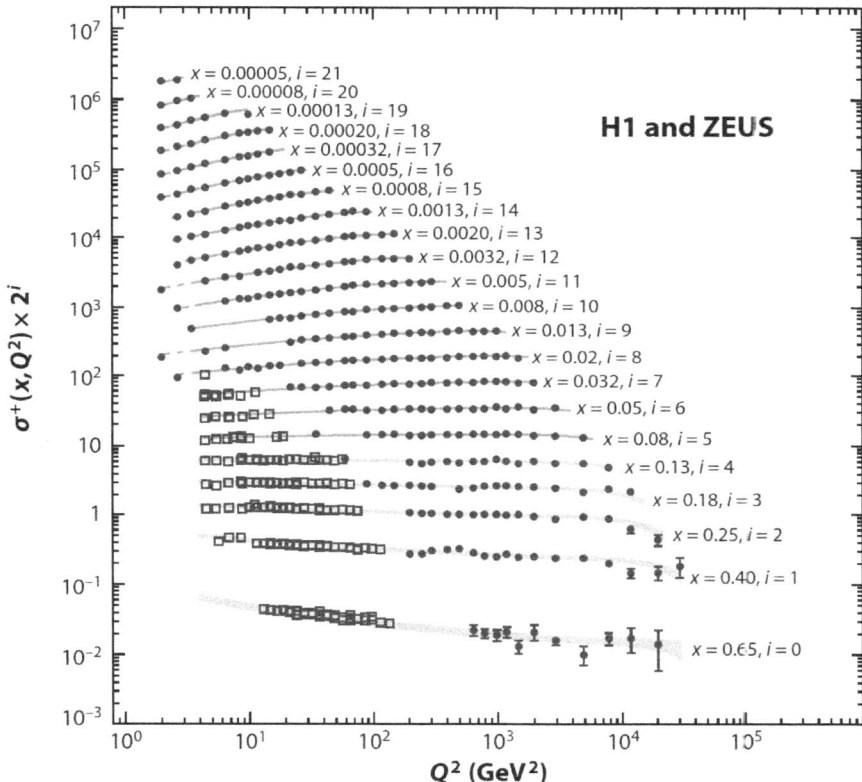

Fig. 6.1: Structure function of the proton measured by the H1 and ZEUS collaborations at HERA as a function of Q^2 in different x bins from Ref. [10]. Q^2 and x are defined in section 1.5. The rise with Q^2 at low x originates from the gluons in the proton.

pulse stretcher ring and Milner began to learn about the unique characteristics of polarized internal gas targets in 1984–1986 as a member of the Polarized Internal Target Working group at CEBAF, led by Roy Holt. In 1987, the new CEBAF Director Herman Grunder reconsidered the accelerator design and this produced the existing superconducting linear accelerator complex.

The HERMES experiment and CEBAF both began data taking in 1995. The CEBAF linear accelerator employs superconducting cavities to accelerate the electron beam, initially to a top energy of 4 GeV, but later operated at somewhat higher energies. HERMES taking data with an electron beam of energy 27.5 GeV, was clearly in the regime which could be interpreted in terms of hard scattering from quarks. HERMES had a significant impact on the shape of the CEBAF scientific program. For example, polarized electron beams were not part of the initial CEBAF scope. HERMES physicists serving on advisory committees strongly argued for this and polarized electrons were soon being developed for CEBAF. One of the major scientific achievements of CEBAF in its first two decades was the successful application of the parton model to relatively low energy (about 5 GeV) electron scattering. In particular, the semi-inclusive and exclusive measurements being pioneered at HERMES at 27 GeV were quickly adopted at CEBAF and provided an important high-energy crosscheck on the CEBAF data. On the other hand, CEBAF could attain far higher luminosity and this was certainly important in making progress.

Most US-based HERMES collaborators were participants in the CEBAF program and there was certainly mutually stimulating intellectual interactions. Further, the Yerevan, Armenia group had a strong presence both at DESY and CEBAF and were involved centrally in the DVCS measurements at both laboratories, as recounted in Chapter 5. Many HERMES-trained physicists found permanent positions at Jefferson Lab, as is described in Chapter 7.

In 2017, Jefferson Lab completed an energy upgrade to 11 GeV electron beams with a new generation of scientific equipment focused on the continued study of the quark and gluon structure of matter using hard electron scattering.

6.5 MIT-Bates Linear Accelerator Center

Richard Milner returned to MIT after spending the years 1995–1997 with his family at DESY working on the HERMES experiment. Milner recalls his last conversation before he departed DESY with Director Bjørn Wiik who advised him "not to disappear into the catacombs of MIT". However, almost immediately upon his return to

MIT, Milner was drawn into a senior leadership role being appointed Director of the Bates Linear Accelerator Center in July 1998.

With the ramp-up of the CEBAF and RHIC flagship facilities, there was pressure to phase out the nuclear physics facility at MIT, which began operation in 1974. However, under the leadership of successive Directors Ernest Moniz and Stanley Kowlaski, Bates focused on realization of a 1-GeV pulse stretcher ring that allowed for polarized internal gas targets. Further, Bates had developed a world-class capability to deliver polarized electron beams with Manouchehr Farkhondeh playing a leadership role.

The small but excellent Bates staff, led by Milner and Associate Director Christoph Tschalär, working with successive LNS Directors Robert Redwine and June Matthews to secure adequate funding, and with strong institutional support from successive Deans of Science Robert Birgeneau and Robert Silbey, was able to carry out in the years 1998 to 2005 the SAMPLE parity violation experiment, the Out-Of-Plane Spectrometer (OOPS) experiment to determine the shape of the proton and the Bates Large Acceptance Spectrometer Toroid (BLAST) internal target program at the South Hall Ring.

BLAST was a technically similar electron scattering experiment to HERMES but at a much lower energy of 850 MeV. The leaders included Ricardo Alarcon from Arizona State University, MIT research scientists Douglas Hasell, Karen Dow, and Michael Kohl, John Calarco from U. New Hampshire, as well as MIT Hadronic Physics Group faculty Haiyan Gao, June Matthews, Robert Redwine and Richard Milner. Jim Kelsey, who as we saw earlier made essential contributions to the design of the HERMES internal target, was now the BLAST project engineer at Bates. Hauke Kolster, who had worked on the HERMES H/D target, together with Bates staff Evgeni Tsentalovich and Ernest Ihloff, brought the BLAST H/D target into regular operation. It took data in the years 2002–2005 and so was able to take advantage of the polarized H/D target developments for HERMES. Because of their large commitments to making BLAST a success, both Milner and Redwine decided, regrettably but necessarily, to end the MIT participation in HERMES in 2000.

In 2006, Bates transitioned to a Research & Engineering Center operated by the MIT Laboratory for Nuclear Science and has continued to provide valuable and unique capabilities to MIT researchers, particularly those that use particle beams. It was successfully directed by Robert Redwine from 2006 until 2018, when the Bates Center became the direct responsibility of the LNS Directorate.

6.6   RHIC: Hot, Dense Matter and the World's First Polarized Proton Collider

The surprising results from EMC in 1987 on the spin structure of the proton gave rise to other important programs. At BNL, the Relativistic Heavy Ion Collider (RHIC), a flagship facility to search for the hot, dense matter created in heavy ion collisions, was on the drawing boards to be built in the tunnel originally constructed for Isabelle. This high energy physics accelerator had been cancelled in 1983 due to problems with developing the superconducting magnets. Since 1982, physicists had been studying the scientific possibilities with polarized protons in a 200-GeV collider and in May 1989, the BNL Directorate set up a task force led by Gerry Bunce and Mike Tannenbaum "to lay out the potential physics program using polarized protons at RHIC".[15]

The Polarized Collider Workshop at Penn State University in November 1990 led to the formation of the RHIC Spin Collaboration, including accelerator physicists, theoretical physicists and experimental physicists with an interest in spin. Gerry Bunce and Mike Tannenbaum from BNL, Bob Jaffe from MIT, and John Collins and Steve Heppelman from Penn State played a leadership role in organizing this important meeting. A proposal was submitted to the BNL HENP Program Advisory Committee in September 1992 for a program of spin physics using the RHIC Collider. This included a detailed conceptual design for the rotators, Siberian snakes and polarimeters which would be necessary to operate RHIC with polarized protons. Further, it contained proposals by the established collider experiments PHENIX and STAR to carry out studies of spin phenomena using these detectors. Finally, the Spin Accelerator Collaboration in the RHIC collider accelerator division was formed under the leadership of Thomas Roser to take responsibility for the design, construction, and installation of accelerator spin components. Further, Japan made major technical and financial contributions that were essential to the realization of the RHIC-spin program.

It was realized that RHIC had the potential to be operated as a polarized proton collider if the spins could be manipulated in the 200-GeV RHIC ring. A scheme involving systems of magnets called *Siberian Snakes* had been proposed in 1974 by Russian theoretical physicists but the idea had never been tested in a large accelerator. A collaboration led by Alan Krisch realized the Siberian Snake experimentally in measurements in the early 1990s at the Indiana University Cyclotron Facility and this established the technology that made possible polarized protons at RHIC.

In 1995, a review of the physics of spin at RHIC by an external review committee, chaired by Charles Prescott from Stanford, stated that "if sensitivities are reached, the results will be profound and form a cornerstone of the theory of hadronic structure." Milner was a member of this committee and recalls traveling from Boston Logan airport to Islip on a small commuter plane, shortly before he and his family left to spend 2 years at DESY. As Milner climbed into the small plane at Boston airport, he was surprised to find Herman Feshbach, esteemed nuclear theorist and MIT faculty colleague, sitting at the back. It turned out that Herman, who was not a member of the review committee, was also traveling to Islip to also attend the review, organized by BNL Director Nick Samios. Herman sat in the room for 2 days as the review proceeded. He did not say one word but carefully listened to the committee discussions. The heavy ion community was nervous about RHIC being used as a polarized proton collider.

In 2000, RHIC started operations and in the first 5 years the priority was to study hot, dense matter in high-energy heavy ion collisions. In 2005, the RHIC experiments announced the discovery of a "perfect liquid", recreating the conditions that previously existed microseconds after the Big Bang. However, funds to operate RHIC as a polarized proton collider were not available to this point. In January 2006, BNL Director Praveen Chaudhari made the stunning announcement that Renaissance Technology, a hedge fund based in

East Setauket, NY, was contributing \$13 million to operate RHIC in 2007. Jim Simons, a member of the BSA Board and President of Renaissance Technologies, an investment management company, initiated and led the drive to raise the money. The 20 week run in 2007 included operation of RHIC as the world's first polarized proton collider. While this generous contribution enabled the start of RHIC-spin, it ended the tenure of BNL Director Praveen Chaudhari, who four weeks later announced that he was stepping down on 30, April 2006.

Since 2007, the year HERMES ceased running, RHIC-spin has evolved into a successful program involving both STAR and PHENIX experiments, that has yielded new insights into the origin of proton spin, as envisaged by its originators. Because of RHIC-spin, we now know that the gluons contribute significantly to the proton spin and that the light polarized sea quarks possess a flavor asymmetry. Further, the successful operation of the world's first polarized proton collider is a major technical milestone on the path to an electron–ion collider, to be discussed below.

6.7   CERN Muon Beam Experiments

Following the discovery of DIS at SLAC in the late 1960s, European physicists came together in the early 1970s to build an experiment to take advantage of the high-energy muon beams to study nucleon structure. Thus, the European Muon Collaboration (EMC) was born.[16] The initial proposal had physicists from CERN, France, Germany, Great Britain, and Italy. Klaus Rith, then at Freiburg U, was a member of the initial EMC collaboration. Between the years 1974 and 1978, when data taking commenced, the muon beam experimental hall and the EMC apparatus were constructed and installed.

Comparing the measured DIS structure functions from iron with those from deuterium led to the startling discovery that the quark momentum distribution in nucleons bound in nuclei was different from that of free nucleons, known as *the EMC effect*. The EMC effect was quickly confirmed in reanalysis of old data and new measurements at SLAC. While the general consensus is that the effect is due to the momentum and binding of the bound nucleon, its definitive understanding is still elusive.

In 1984 and 1985, polarized targets were installed and the DIS spin-dependent structure functions measured at lower x than the first measurements at SLAC, which were made at high x. Vernon Hughes and his group from Yale was a central figure in this. By 1986 at the collaboration meeting in Oxford it was realized that the polarized structure functions were not behaving as expected. Thus, was born what was called at the time *the spin crisis*. This was reported in the US at the SLAC workshop by EMC spokesman Terry Sloan (see earlier discussion).

After the EMC experiment completed data taking in 1985, a new collaboration formed which became known as the New Muon Collaboration (NMC). This collaboration continued to use the EMC Forward Spectrometer for further studies of deep inelastic scattering in both free nucleons (hydrogen and deuterium) as well as an extensive range of nuclear targets to study the EMC effect. After NMC had finished data taking another new collaboration was formed to study the spin structure of the nucleon. This became known as the Spin Muon Collaboration (SMC). The SMC took data from 1991 to 1999 and used both polarized protons and deuterons to compare the spin structure of the proton and neutron. Together with HERMES, SMC provided key experimental data on DIS and SIDIS that shaped the development of the 21st century picture of the proton.

Finally, the apparatus and infrastructure were taken over by the Common Muon and Proton Apparatus for Structure and Spectroscopy (COMPASS) Collaboration in 2000 to continue both muon and hadron physics in the North Area beam line. COMPASS completed data taking in 2018. At present, another evolution of the muon beam experimental program, called COMPASS++/AMBER, is in preparation with plans to begin data taking in 2022.

The CERN muon program has been a hugely important source of insight into the fundamental quark and gluon structure of matter over almost half a century.

6.8   Fermilab: Annihilating Quarks and Antiquarks

As we have explained earlier in this book, electron scattering from the proton provides the most direct experimental means to access the electrically charged quark and, indirectly the gluon constituents. However, it is possible to annihilate the quark from a projectile with the antiquark of the target to produce a photon which decays into a lepton–anti-lepton pair. This *Drell-Yan process* also gives direct access to the quark and antiquark distributions and the dilepton pair, e.g. muon–anti-muon, are straightforwardly detectable. In 1992, Gerald (Gerry) Garvey and Jen-Chieh Peng and colleagues initiated the E-866 experiment that has shown a remarkable difference in the distributions of anti-*up* quarks and anti-*down* quarks in the proton. This flavor asymmetry in the sea quarks is consistent with a significant role by clouds of mesons in proton structure. Recently, the E-906/SeaQuest experiment has pursued more precise studies of this effect and a polarized experiment to probe the polarized sea quarks is planned.

6.9   LHCspin: A Storage Cell in the 7-TeV Beam

As discussed in Chapter 3, it was possible to realize the HERMES experiment at the HERA collider because essential features had been foreseen in the machine design, the most important being longitudinal electron ($e^{\pm}$) polarization at the IP's, based on the Sokolov–Ternov effect, i.e. polarizing the beam at the final energy. There were also extensive studies of implementing a 920-GeV spin-polarized proton beam at HERA by Alan Krisch, Desmond Barber and their colleagues. As no process is known to polarize such a beam at its final energy, it would be necessary to accelerate a polarized beam from the source all the way up to the final energy, crossing many depolarizing resonances, as realized at RHIC-spin (see Section 6.3). At HERA, there was insufficient space in the accelerator for the required number of Siberian Snakes necessary to maintain the proton polarization.

The same constraint exists at the Large Hadron Collider (LHC) facility at CERN with its two counter-rotating proton beams, each with an energy of 7.0 TeV (Tera = T = 10^{12}). Practically, it is impossible to produce and maintain polarization in these beams. However, it is possible to measure spin-dependent observables at the LHC if the unpolarized 7 TeV proton beam is scattered from a polarized HERMES-type internal gas target. In particular, the angular distribution of reaction products from a transversely polarized proton target yields so-called single-spin asymmetries (discussed earlier in Section 1.5), which were successfully measured at HERMES in 2002–2005.

In 2012, a first report led by Stanley Brodsky has highlighted the potential of fixed-target spin experiments using the LHC beams.[17] In 2015 at the PSTP workshop in Bochum (Germany), Steffens presented estimations on the performance of a HERMES-type hydrogen target in the LHC beam.[18] It became obvious that a storage cell target is the optimum choice for a LHC gas target, both polarized (H, D, ^{3}He) and unpolarized (inert gases like H_2, He, N_2, Ne, etc.). Following intense discussions within the AFTER@LHC initiative,[19] a study group of former HERMES experimentalists (Ferrara, Frascati, Petersburg, Erlangen), led by Pasquale Di Nezza, began to explore the addition of an unpolarized and a spin-polarized gas target upstream of VELO, the Vertex Locator of the LHCb detector. The project, called LHCspin, is a continuation of SMOG (System for Measuring the Overlap with Gas) used at LHCb to calibrate the luminosity of the colliding bunches. In addition, SMOG enabled, for the first time, fixed-target (FT) measurements at the LHC, e.g. p (7 TeV) + He-4 (at rest), using the LHCb detector. These measurements have been of special interest for the production of antiprotons in interstellar space and the interpretation of space-based experiments like the Alpha Magnetic Spectrometer on the International Space Station.

As a first step towards a polarized storage cell target, the addition of SMOG2 to the VELO detector has been proposed and accepted, consisting of a storage cell (see Section 3.1) and a gas injection system. The goal is two-fold: (i) to improve the conditions for fixed-target measurements by a localized pressure bump, and (ii) to study the effect of a dense gas target on the stability of the LHC beam. Installation is scheduled for the long shutdown (2019–2021) with the option of first measurements in 2021. In parallel, a conceptual design study of a compact HERMES-type gas target is underway in the course of the LHCspin project.

6.10   Towards the Electron–Ion Collider

In the 1990s, with the new facilities CEBAF and HERMES turning on and SMC taking data at CERN, there was a consensus in the QCD community in Europe and N. America that the comprehensive study of the fundamental quark and gluon structure of the visible matter in the universe would require a new electron accelerator specifically designed for this purpose. QCD's evolution equations demanded an accelerator that would span a large range in Q^2. The ELFE project proposed a linear accelerator of energy about 20 GeV and this was considered in Europe in the late 1990's. However, ultimately the science demanded a bolder vision than that provided by factors of two increase in electron linac beam energy.

The HERA collider, although proposed as an accelerator to look for substructure of the quarks and leptons, as we have seen at the beginning of this chapter, had provided important new insight into the QCD structure of the proton. Thus, it was realized that an electron–ion collider with both polarized nucleons and nuclei and possessing a higher luminosity than achieved at HERA, would be the ideal machine to continue the study of the fundamental structure of matter into the 21st century. In a sense, EIC combines the pure spin-dependent lepton–nucleon scattering capabilities of HERMES with the collider experiments ZEUS and H1 with the important addition of nuclear ion beams.

In March 1997, a joint DESY/GSI/NuPECC meeting took place at Seeheim, Germany to consider the long range perspectives of electron-nucleon/nucleus scattering in Europe. Three projects were presented: electron-nucleus collisions in HERA, a new electron-nucleon/nucleus collider with center-of-mass energy in the range 20–30 GeV, and ELFE@DESY. Ultimately, none of these projects came to fruition in Europe but they sowed fertile seeds that eventually sprouted as the electron–ion collider project in the US.

In the last decade of the 20th century and first decade of the 21st century, theorists played a crucial role in inventing a new language in which to describe the physics of the fundamental structure of matter. The realization that one could access the transverse structure of the target in high energy electron scattering revolutionized the thinking about a future accelerator. New observables were identified that demanded unique particle identification and kinematics as well as high collision luminosity that drove the specifications of the EIC. An entire new effective field theory, called the *color glass condensate* was developed to describe the QCD structure of the proton or nucleus at low x. This provided a basis for detailed study of elusive dense gluon configurations.

Another aspect of future planning was that university-based medium energy facilities in the US were being phased out and the physicists at these institutions were forced to seriously consider the future of the field. The NSF-funded IUCF accelerator was closing and John Cameron, Tim Londergan and others there concluded that a low-energy electron-proton collider (EPIC) offered great scientific potential. IUCF hosted a workshop on EPIC in 1999. At the DOE-funded MIT-Bates accelerator center, the last phase of polarized internal gas targets with the BLAST detector was underway, and Manouchehr Farkhondeh, Christoph Tschalär, Richard Milner and others also concluded in the late 1990s that an electron-ion collider was the most promising accelerator configuration.

Independently at BNL, Abhay Deshpande, Gerry Garvey, Vernon Hughes, Larry McLerran, Peter Paul, Raju Venugopalan and others were working on an electron–ion collider based on the RHIC complex, called eRHIC. They held their first eRHIC workshop in December 1999 and a second at Yale in April 2000, even before RHIC started operation. MIT hosted a successor workshop to the 1999 meeting at IUCF on EIC in September 2000 and both Frank Wilczek and Bob Jaffe gave supporting scientific talks. This meeting brought together the IUCF, MIT and eRHIC/BNL colleagues and it was decided to work together to make the strongest case at the upcoming 2002 Long Range Planning Exercise.

The 2002 US Nuclear Physics Long Range Planning Exercise was a significant early milestone in the path to EIC. This is a regular exercise held about every 6 years where the plans are formulated for about the next decade. The major decisions are made by a group of about 70 physicists who meet typically in a secluded location for about 3 days. In March 2001, this group led by James Symons met near Santa Fe, New Mexico and Richard Milner was the lead proponent of an Electron–Ion Collider, eRHIC. Abhay Deshpande made a presentation "Physics Case for eRHIC" and there were also presentations on CEBAF24, an energy upgrade to 24 GeV. Abhay Deshpande recalls CEBAF accelerator physicist Swapan Chattopadhyay remarking to his colleague Lia Merminga "Why not turn CEBAF12 into an injector for a future collider?" This may have been the beginning of the CEBAF-based EIC design. Gerry Garvey from Los Alamos was eloquent in making the science case for eRHIC. Prominent nuclear physics theorist Vijay Pandharipande spoke up in support of consideration of eRHIC. In the same meeting, BNL was pushing RHIC II and Jefferson Lab was pushing the CEBAF energy upgrade. However, early in the meeting a secret vote to rank scientifically about a dozen new initiatives was carried out. eRHIC came out near the top ahead of either RHIC II and the CEBAF upgrade. While, eRHIC was not in the end a high priority of the 2002 long range plan, the fact that the science had strong support across the US nuclear physics community was established. Subsequently, eRHIC was featured in 2003 in DOE's "Facilities for the Future", a 20 year outlook for the future from the DOE Office of Science.

In April 2002, Paola Ferretti Dalpiaz from Ferrara, Enzo De Sanctis from Frascati and Wolf-Dieter Nowak from Zeuthen chaired the European Workshop on the QCD Structure of the Nucleon at the Castello Estense in Ferrara, Italy. This produced the Declaration of Ferrara, which eloquently laid out the case for a next-generation electron accelerator to pursue the study of QCD.

Meanwhile, back at DESY, then the world center of high-energy electron–proton scattering, there was a concerted effort to make the case for a post-2007 HERA-III era, where nuclear beams would be accelerated and collided with the 27 GeV electron beam and the interactions studied in the existing H1 and ZEUS detectors. Ferdinand Willeke led the accelerator effort to collide nuclei up to calcium and the estimated cost was DM 50 million. However, the German high energy physics community had little interest in HERA-III and rather chose to focus on the LHC at CERN. Thus, HERA running came to an end in 2007. Subsequently, several of the leaders of the HERA-III project, including Max Klein, originated LHeC, the high-energy electron-ion collider project at CERN.

After the 2002 Long Range Planning Exercise, there was a strong push from BNL management and their Program Advisory Committee to get the broader electromagnetic nuclear physics community to participate in eRHIC. Both Deshpande and Milner were regularly confronted with: "if this is so important for QCD why isn't Jefferson Lab interested?" Well, this happened quickly, maybe in a manner which they had not intended. Rolf Ent from Jefferson Lab played a key role in originating and leading an effort at Jefferson Lab in EIC at a time when the lab was still making the case for the CEBAF energy upgrade. The accelerator physicists at JLab quickly became interested in a CEBAF-based EIC design, beginning in late 2001. Lia Merminga was an originator and an effective group formed including Swapan Chattopadhyay, Yaroslav Derbenev, Andrew Hutton, Geoff Krafft, Matt Poelker, Yuhong Zhang and colleagues. Thus, the second EIC Workshop took place at Jefferson Lab in March 2004 and the figure-eight ring-ring design, originated by Yaroslav Derbenev, was introduced. This concept, based around the existing CEBAF, was called ELIC, and remained a stable and effective design that was further optimized over almost two decades.

In 2004, the first EIC accelerator conceptual design was developed by a Bates–BNL collaboration, the so-called *Zeroth order Ring-Ring eRHIC Design*, coordinated by Manouchehr Farkhondeh and Vadim Ptitsyn. The maximum luminosity in this design was 10^{32} cm^{-2}s^{-1}. Christoph Tschalär from MIT-Bates and Desmond Barber from DESY brought their considerable expertise from commissioning and operating polarized electron beam storage rings in their respective laboratories.

After the development of the initial eRHIC ring-ring concept in 2004, the BNL eRHIC machine design group focused for about a decade on an electron linac colliding with the RHIC ion beam. While this concept had the possibility of attaining higher collision luminosity than the ring-ring concept, it demanded enormously high currents of polarized electrons as well as very ambitious hadron cooling mechanisms. This drove ambitious R&D for over a decade at BNL. Ultimately, it became clear that a ring-ring concept based around the existing RHIC accelerator, as conceived initially in 2004, was less risky and the most realistic option to realize EIC at BNL.

In April 2007, in an EIC meeting at MIT, the EIC Collaboration (EICC) was formed with three goals: (1) to develop the most compelling science case for a high luminosity, high energy electron-ion collider (EIC) independent of where it may be sited; (2) to work with accelerator physicists, especially at BNL and JLab, to develop the optimal EIC accelerator design; (3) to design and realize through a R&D program the most optimal suite of detectors for the EIC. The following EICC Steering Committee was established: Allen Caldwell (MPI Munich), Abhay Deshpande (Stony Brook) (Co-Chair), Rolf Ent (JLab), Gerry Garvey (LANL), Emlyn Hughes (Caltech), Ken'ichi Imai (Kyoto U.), Peter Jacobs (LBNL), Lia Merminga (JLab), Richard Milner (MIT) (Co-Chair), Peter Paul (BNL), Jen-Chieh Peng (U. Illinois), and Thomas Roser (BNL). EICC was active through the 2010 Long Range Planning Exercise and held meetings at BNL, JLab, U Maryland, MIT, Washington DC, Stony Brook U, Hampton U, GSI, LBNL, U. Michigan, Trento, INT, and Catholic U.

In April 2007, the resolution meeting of the US Long Range Plan took place in Galveston, Texas and discussions were led by Robert Tribble. This was preceded by Town Meetings in the different subfields. The principal EIC proponents at the Galveston meeting were Abhay Deshpande, Rolf Ent, Richard Milner and Thomas Ullrich. A strong recommendation to pursue EIC R&D together with a dedicated chapter on the science in the report meant that the forward momentum for the EIC project continued. Steven Vigdor played a significant role in writing the most effective case for EIC in the 2007 long range plan document.

Following the 2007 US Long Range planning exercise, an intensive ten-week workshop took place in fall 2010 at the Institute for Nuclear Theory at the University of Washington Seattle. The purpose was to bring together the world's experts in QCD to make a comprehensive and broad study of the EIC science case with the aim of then writing a succinct and compelling summary paper for the subsequent US long range planning exercise. Thus, a 538 page proceedings from the INT workshop was published in August 2011. Subsequently, Brookhaven National Laboratory and Jefferson Laboratory, with Steven Vigdor (and later Berndt Mueller) and Bob McKeown together playing an important leadership role, led the writing of the EIC White Paper, which was first published in 2012, and revised in 2014. Abhay Deshpande, Jian-Wei Qiu and Zein-Eddine Meziani were the lead editors of the EIC White Paper. This EIC White Paper presented the science case which was reviewed in the 2015 US Long Range planning exercise.

In September 2014, a town meeting of the entire US QCD community took place at Temple University in Philadelphia. The EIC was endorsed unanimously as the highest priority for new construction, after existing commitments were fulfilled. In April 2015, the Long Range Planning resolution meeting took place at Kitty Hawk, North Carolina led by Donald Geesaman. It unanimously endorsed the Electron-Ion Collider as the next facility to study QCD, after completion of existing commitments. Thus, an initiative which began in the mid-1990s by those intent on understanding the fundamental quark and gluon structure of matter using lepton scattering, had finally become the top priority for future construction for the US nuclear physics community.

In 2017 and 2018, a committee appointed by the US National Academy of Sciences and Engineering evaluated the science case for EIC. The committee, chaired by Ani Aprahamian and Gordon Baym, released a report in July 2018 that favorably endorsed the science case for EIC and emphasized the strategic importance for the US of the accelerator science technology associated with EIC realization.

Further, in 2016 an EIC User Group was founded with the aim of promoting the science and technical case for EIC, independent of the siting. By 2021, this had over 1200 Ph.D. members from more than 250 institutions in 34 countries across the globe.

In January 2020, the US Department of Energy (DOE) announced the official launch of the EIC project and that Brookhaven National Laboratory would be the site for the new accelerator. The DOE Office of Nuclear Physics has carefully steered the EIC project over two decades with Dennis Kovar, Timothy Hallman, Jehanne Gillo and Manouchehr Farkhondeh providing important leadership.

HERMES physicists over the decades have played an active role in charting the long and winding path towards a next-generation accelerator for the study of the fundamental quark and gluon structure of matter, which has led us to EIC. Excellent examples include Elke Aschenauer and her group at BNL, who have led the detailed design of the optimal EIC detector. At Jefferson Lab, Markus Diefenthaler coordinates the EIC software development. Many others, who have spent time in the HERA East Hall, are active in the realization of EIC. Finally, Ferdinand Willeke, the maestro of HERA operations and in whose company the authors spent many hours in meetings at DESY, is leading the EIC accelerator effort from BNL.

Chapter 7

HERMES Collaboration

7.1 Introduction

The HERMES experiment was carried out by a collaboration that was initiated at the meeting in Rockport, Maine in May 1988. At the time of writing, over 32 years later, this collaboration still is active. Here, we describe the collaboration, the contributions of the different institutions, the leadership and give an impression of the regular activities, particularly in the data taking phase from 1995 to 2007.

7.2 Contributions of the Different Institutes

In Chapter 3, we have described how the HERMES collaboration came about. There were 16 founding institutions from Canada, Germany, Italy and the United States that proposed the HERMES experiment in January 1990. By mid-July 1991, discussions with additional collaborators from DESY, DESY-Zeuthen, NIKHEF, Yerevan, St. Petersburg, Frascati and Rome were in progress.

In 1992, Klaus Rith moved to the University of Erlangen to establish a large group that formed the cornerstone of the HERMES experiment for the duration of its existence. In 1995, Erhard Steffens left MPI-K Heidelberg and joined the Erlangen group. Klaus, as Co-spokesperson of HERMES, put a lot of effort into finding strong groups which could help to cover all the important responsibilities within the experiment.

In 1992, the University of Colorado joined the HERMES collaboration as well as Rome, St. Petersburg, and Zeuthen. In 1993 Frascati, NIKHEF and Yerevan joined and by then Los Alamos, Stanford, Torino and College of William and Mary had left. In 1994, the HERMES collaboration extended to Asia when the Tokyo Inst. of Technology joined. In 1995, Gent and Freiburg joined HERMES. In 1996, Ferrara and Michigan became collaborators. In 1998, the theory group of Andreas Schäfer at Regensburg joined the HERMES collaboration. In 2000, the Caltech and MIT groups left the HERMES collaboration due to their increasing commitments in other experiments in the US. In 2001, groups from Beijing, Glasgow and Warsaw joined HERMES. Throughout the lifetime of the HERMES experiment, 38 institutions were active members.

Table 7.1 lists the HERMES institutions and a summary of each institution's responsibilities, as defined in the Technical Design Report of July 1993. More than 135 Ph.D. theses have been written on HERMES data.

Responsibilities for major components and important activities were shared among several institutions. These included

Polarized H/D Target
Erlangen, Heidelberg, Liverpool, Madison, Marburg, Munich
Polarized ^{3}He Target
Caltech, MIT
Calorimeter
Amsterdam/NIKHEF, Frascati, New Mexico State U, Yerevan
Transition Radiation Detector
Alberta, Simon Fraser U, TRIUMF
Magnet
Argonne, DESY, Heidelberg, Rome
Ring Imaging Cherenkov Detector
Argonne, Bari, Frascati, Gent
Analysis of Data
Common task of the institutes

Table 7.1: Summary of the major contributions to the hardware of the
HERMES experiment by the collaborating institutions.

Institution	Responsibility
Alberta	Transition Radiation Detector, trigger
Amsterdam/NIKHEF	Calorimeter, microstrip gas detectors
Argonne	Cerenkov, RICH, Laser Driven Source, magnet
Bari	RICH, recoil detector
Caltech	Hodoscopes, polarized ^{3}He target, calorimeter
Colorado	Forward drift chambers
Beijing	Atomic Beam Source
DESY	Infrastructure in East Hall
Dubna	Tracking chambers
Erlangen	Polarized H/D target, drift chambers, lumi. monitor
Ferrara	Polarized H/D target
Frascati	Calorimeter, RICH, recoil det., slow cntrls.
Freiburg	Longitudinal beam polarimeter
Gent	Radiative corrections, RICH, hodoscopes
Glasgow	Recoil detector
Heidelberg	On-line DAQ, polarized H/D target, Magnet
Illinois	Monte-Carlo
Liverpool	Polarized H/D target
Madison	Polarized H/D target
Marburg	Polarized H/D target
Michigan	Longitudinal polarimeter
MIT	Vacuum system, unpolarized gas system, polarized ^{3}He target
Moscow	Luminosity monitor
Munich	Polarized H/D target
New Mexico State	Trigger electronics, veto hodoscope, magnet
St. Petersburg	HERMES magnet, Magnet chambers
Protvino	Trigger
Rome	Magnet chambers
Simon Fraser U	Transition Rad. Detector
Tokyo	Alignment system
TRIUMF	Transition Rad. Detector
Yerevan	Calorimeter
Zeuthen	Drift chambers

7.3 Leading Contributors to HERMES

Here, we identify HERMES colleagues whose contributions deserve specific mention because of their leadership and essential contributions to the HERMES experiment (see Collaboration picture in Fig. 7.1). Of course, this is a personal choice by the two co-authors and in no way diminishes the substantial contributions of other colleagues.

Fig. 7.1: The HERMES Collaboration in front of their experiment in the HERA East Hall. The picture was taken on 19 January 1995, during a collaboration meeting at DESY. Two weeks later, on 6 February, the experiment was rolled into beam position (*Source*: DESY).

Klaus Rith was the single individual most responsible for the success of the HERMES experiment. He was a co-founder in 1987–1988 and a leader in assembling the international collaboration. His group was the largest single collaborating institution with 24 graduate students from Erlangen writing Ph.D. theses on HERMES. Klaus and his group also put significant effort into the development of the laser driven source of hydrogen and deuterium. The HERMES analysis framework was shaped largely by Klaus' group. Further, the Erlangen group has been very active in HERMES up until the present day, even though Klaus formally retired from the Erlangen faculty in 2008. After his retirement, Klaus served for another 3 years as HERMES spokesperson (Fig. 7.2).

Bogdan Povh was a Director of MPI-K, Heidelberg. His research was concentrated at CERN, on Hyper-Nuclei at the 28-GeV Proton Synchrotron, and at the LEAR facility with antiproton beams. He initiated the FILTEX proposal (see Section 3.3.1) leading to a novel target development. With the appointment of Klaus Rith, he created a strong group at MPI-K and had a key role in the realization of the HERMES experiment (Fig. 7.3).

Fig. 7.2: Klaus Rith, co-originator of the HERMES experiment as a member of Povh's MPI-K department, who built up one of the strongest HERMES groups at the University of Erlangen.

Fig. 7.3:   Bogdan Povh, Director at the MPI-K Heidelberg with excellent ties
to CERN and DESY, and one of the key early supporters of the HERMES
experiment (*Source*: B. Povh).

Michael Düren played an important role in the design of the
HERMES experiment, served as Analysis Coordinator in the early
years of HERMES and oversaw the development of the analysis
framework as a member of Klaus Rith's group at Erlangen. Subsequently, he moved to the University of Giessen, where he built a
strong research group (Fig. 7.4).

Wolfgang Wander, in his Ph.D. work for the University of
Erlangen, played a central role in the development of the HERMES
track reconstruction software. Already as a student, he helped to set
up a readout system for the transverse electron polarimeter (TPOL,
see Section 3.7) which allowed for remote control of the TPOL via
internet e.g. from Erlangen.

Antje Bruell from the University of Heidelberg, was a leader
in the HERMES analysis effort in the early, formative years. She
served as Deputy Spokesperson during the first years of data taking
and subsequently took positions at MIT and Jefferson Lab. Later,
Naomi Makins, from the University of Illinois, played a leading role
in the HERMES analysis effort, in the ground breaking years of
transverse polarized target running. Naomi and her group developed

Fig. 7.4: Michael Düren, one of the designers of the HERMES experiment when working in Rith's groups at Heidelberg and Erlangen; now at University of Giessen.

Fig. 7.5: Naomi Makins from the University of Illinois at Urbana-Champaign, and Delia Hasch, from Frascati, both leaders of the HERMES experiment (*Source*: HERMES Coll.).

effective visualizations of the new fragmentation processes which involved transverse momentum of the quarks in the nucleon. These served to facilitate discussion between theorists and experimentalists in developing the modern picture of the structure of the proton (Fig. 7.5).

Fig. 7.6:   Harold Jackson, from Argonne National Laboratory, one of the leaders of the HERMES experiment.

Harold (Hal) Jackson was one of the founding members of HERMES and led a sizable and effective group from Argonne National Laboratory over the lifetime of HERMES. Hal and his group principally focused on hadron detection and played a leading role in the Cherenkov and subsequent RICH detectors, that were so important for the HERMES scientific program. Hal was an excellent physicist with an easygoing personality who often helped to achieve consensus from contentious discussions within the collaboration. Unfortunately, Hal passed away in 2019 (Fig. 7.6).

The Canadian groups were founding members of the HERMES collaboration and played a leading role from the beginning. Their involvement was initiated by Otto Häusser, the outstanding Canadian physicist from Simon Fraser University and TRIUMF. The Canadian groups acted as an effective collective force based around the TRIUMF laboratory in Vancouver, Canada. They were responsible for the crucial Transition Radiation Detector necessary for good particle identification. Regrettably, Otto passed away in 1998 at a relatively young age. Mike Vetterli took over the leadership role and made essential contributions to the formation of HERMES and to the commissioning and early years of running (Fig. 7.7). Further, Andy Miller from TRIUMF is a superb physicist both

Fig. 7.7: Michel (Mike) Vetterli from SFU/TRIUMF and Ed Kinney from the University of Colorado at Boulder, both leaders of the HERMES experiment (*Source*: HERMES Coll.).

technically and scientifically with the ability to understand very complicated problems. He was a strong force within the leadership of the experiment throughout its lifetime. A notable achievement was his leadership role in figuring out a subtle effect involving radiation corrections from nuclei, which had attracted attention outside the collaboration. Andy showed that what was starting to be interpreted as a novel physics effect was, in fact, an instrumental effect that originated from radiation in the forward region of the HERMES spectrometer (Fig. 4.8).

Special credit for the success of the experiment is due to the two target groups which had the difficult task to apply the novel technology of polarized gas targets in the challenging environment of a high-energy electron ring for the first time:

(1) the Caltech-MIT group for the Helium-3 target which ran smoothly during the first year 1995 of HERMES operation (see Section 4.6.1);

(2) the   Heidelberg-Erlangen-Ferrara-Liverpool-Madison-Marburg-München group for the H/D target which worked for 10 years within the experiment in different configurations (see Sections 4.6.2 and 4.7.2).

For the H/D target, the continuous support by the participating institutes during the very long period from 1985, when the FILTEX development started, to 2005, when the H/D target was replaced by the recoil detector, was essential for the final success. In addition to the early supporters Bogdan Povh, Willy Haeberli and Klaus Rith, and leaders of the target group Geoff Court and Paolo Lenisa, important contributors with their strong groups were Gerhard Graw and Peter Schiemenz (München), unfortunately, Peter passed away in 1999; Paola Ferretti-Dalpiaz (Ferrara, see below) and Dieter Fick, Steffens' long-standing cooperation partner, first as group leader in the department of Peter Brix at MPI-K, then as Chair Professor at U. Marburg. Key innovations to target technology and analysis were contributed, among several others, by Hans-Günter Gaul and Friedemann Stock (Heidelberg), Christian Baumgarten (U. München), Thomas Wise (U. Wisconsin), Wolfgang Korsch (U. Marburg), and Norbert Koch and Alexander Nass (U. Erlangen). More than 15 Ph.D. thesis' and twice as many Diploma thesis' were written during the development and running of the H/D target.

For the polarized ^{3}He target used in the initial data taking in 1995, the leadership and dedication of Stephen Pate and Laird Kramer from MIT to maintaining the target at DESY was essential. MIT engineer Jim Kelsey also played a crucial role in ensuring that the HERMES experiment had a polarized gas target with full differential pumping system in its initial year of running. Graduate student Dirk De Schepper was another key member of the MIT team. The Caltech group's focus in 1995 was on the target polarimetry and Mark Pitt, Bob Carr and graduate student Andrea Dvoredsky were leaders of this effort at DESY.

In 1996, a group from Ferrara, under the leadership of Paola Ferretti Dalpiaz, joined HERMES. They became major contributors to the successful operation of the H/D target for the second phase of the HERMES experiment. In 1998, Paolo Lenisa, an atomic physicist who worked before at the Heidelberg storage ring (TSR), became target coordinator and steered it successfully throughout the deuteron and transverse proton measurements. Presently he is, as leader of a large group at INFN Ferrara, very active in spin physics at Forschungszentrum Jülich and the LHC/CERN (see Section 6.6).

Johannes (Jo) van den Brand, educated at NIKHEF, was a post-doc with Milner at MIT in 1989–1991, and became a leader in the HERMES experiment, initially at the University of Wisconsin-Madison, and after 1993 at NIKHEF. Jo was trained as a mechanical engineer before pursing a Ph.D. in physics and is an experimental physicist of enormous technical prowess. Jo was HERMES Spokesperson in 1994–1995 during the intensive installation and commissioning of HERMES. Subsequently, Jo led a program of experiments at the Amsterdam Pulse Stretcher Ring, and was active in the LHCb experiment at the LHC. His most recent research focus has been in gravitational wave detection and he served as Spokesperson of the Virgo Gravitational Wave Collaboration from 2017 to 2020 during the exciting period of scientific discovery by the joint LIGO-Virgo collaboration.

Elke Aschenauer made essential contributions in leading the HERMES experiment in the productive HERA-II years of 2001–2007. As a NIKHEF post-doc she was active in the commissioning and first running in 1995. In September 1996, she joined HERMES full-time and played a very active role in leading the experiment until the end of 2006. She served as Deputy Spokesperson from 2000 to 2003 and followed this with 3.5 years as HERMES Spokesperson. She oversaw the design, installation, commissioning and running with the recoil detector. In 2009, Elke moved to BNL to take a leadership role in the RHIC-spin experiment. Subsequently, she has built an effective group at BNL that plays a leading role in the development of the EIC physics case and the design of EIC detectors (Fig. 7.8).

The Yerevan Physics Group from Armenia played an important role in the HERMES experiment. They had a long tradition of

Fig. 7.8:   Elke Aschenauer, now at Brookhaven National Laboratory, one of the leaders of the HERMES experiment.

research in electro-nuclear physics and had built a 6-GeV electron synchrotron accelerator in the 1960s, in close collaboration with the DESY laboratory where a similar machine was built. With the dissolution of the Soviet Union in 1991, Armenian physicists were attracted to the West and DESY supported their local expenses in Hamburg. Thus, as described earlier in Section 4.2, Klaus Rith visited Yerevan in summer 1991 and negotiated that a sizable Yerevan group would join the HERMES collaboration. In 1994–1995, they played a central role in the successful assembly, testing, installation and commissioning of the electromagnetic calorimeter. Subsequently, Yerevan was a potent group in the analysis of HERMES data. Moscow Amarian was the principal driver to extract DVCS data for the first time, as described earlier in Chapter 5. Harut Avakian was a principal in the extraction of single-spin asymmetries (SSA) which had a big scientific impact, particularly with the transversely polarized target (Fig. 7.9).

The Italian institutes in HERMES from Bari, Ferrara, Frascati, and Rome made important contributions. At the beginning phase, Frascati played a leading role in the realization of the electromagnetic

Fig. 7.9: Robert Avakian from Yerevan, his son Harut Avakian from Jefferson Lab and Enzo De Sanctis, from Frascati, deep in conversation at the HERMES end of data taking celebration in June 2007 (*Source*: HERMES Coll.).

calorimeter and Rome, together with St. Petersburg, were responsible for the tracking chambers in the magnet. Enzo De Sanctis and Salvatore Frullani brought great experience and provided wise and effective leadership of these important components of the HERMES spectrometer. Bari and Frascati contributed to the realization of the RICH detector. Ferrara was active within the target group, see above.

The group of Gent U., lead by Dirk Ryckbosch, contributed strongly to the CHARM upgrade, in particular to the RICH detector and by work on radiative corrections. Dirk served as spokesperson 2001–2003.

Gunar Schnell started at HERMES as a student of Gary Kyle (NMSU) in 1996 and wrote a Ph.D. thesis on Λ polarization in 1999. After completing his military service in the German army, he joined the DESY-Zeuthen group in HERMES in 2000. He has been a major force post-2007 in driving the analysis of HERMES data. In 2011, he took a position in Bilboa, Spain, from where he leads the HERMES analysis effort (Fig. 7.10).

In addition to the HERMES collaboration, it is important to recognize the important contributions of key DESY personnel that were essential to the success of HERMES. The HERA machine group played a key role in the design and realization of the successful HERMES experiment. Gus Voss and Bjørn Wiik were the

Fig. 7.10:   Gunar Schnell, now at the University of Bilboa, the longest-serving Spokesperson of HERMES. He managed to continue the editing of HERMES publications, more than ten years after the end of data taking (*Source*: HERMES Coll.).

principal architects of the HERA collider and played an important role in the technical considerations related to integration of the HERMES experiment in the HERA accelerator (Fig. 7.11). Reinhard Brinkmann was a leader in the accelerator physics group and Ferdinand Willeke led HERA operations and chaired the weekly meeting where the DESY machine group and the experiments discussed machine performance and improvements. Desmond Barber was the DESY physicist who studied the electron polarization and developed important simulations that were valuable in optimizing the polarization value, once a finite signal was detected.

Finally, HERMES would never have been realized without the outstanding and enlightened leadership of the DESY Directorate over three decades. At the dawn of HERMES, Volker Soergel (Fig. 7.12) and Paul Söding were open to serious consideration of a technically innovative experiment by a collaboration of nuclear physicists that addressed physics quite different to that for which HERA was constructed. Soergel was DESY Director from 1981 to 1993 and oversaw the birth and realization of the HERA collider. In this time, DESY pioneered a new model for international collaboration on large scientific projects. While cautious about granting full approval until

Fig. 7.11: The renowned accelerator physicist Gustav-Adolf (Gus) Voss (1926–2013), recipient, among other distinctions, of the AIP Tate medal 2009 for promoting international physics, famous for his work on low-beta sections, and responsible for the design of PETRA and the HERA electron ring. He was affectionately known as the "Lord of the Rings". He encouraged cooperation with East European scientists and promoted the SESAME project in Jordan (photo courtesy of DESY).

the polarized electron beam was established in HERA, the DESY Directorate supported the efforts of the HERMES collaboration to secure funding. Subsequently, DESY Director Bjørn Wiik from 1993 to 1999 oversaw the successful construction, installation and operation of HERMES (Fig. 7.12). During this period, DESY Research Director Albrecht Wagner played a key leadership role, as described earlier in Chapter 4. Following Wiik's unfortunate accidental death in 1999, Wagner became DESY Director through the end of the running of HERA until 2009. Rolf Heuer was DESY Research Director from 2004 to 2009, when he became Director General of CERN. In 2009, Joachim Mnich assumed the position of Director for Particle and Astroparticle Physics at DESY. The success of HERMES would not have been possible without the strong, sustained support of the DESY laboratory leadership over several decades.

Fig. 7.12:   Successive DESY Directors Volker Soergel (1981–1993) and Bjørn
Wiik (1993–1999). Credit: Deutsches Elektronen Synchrotron (DESY), Hamburg,
courtesy of AIP Emilio Segrè (*Source*: Visual Archives, Physics Today Collection).

Table 7.2 lists the HERMES Spokespersons, Deputies, Technical
Coordinators, and Analysis Coordinators from 1988 to the present.

7.4   HERMES Spanned the Pre-email Era, through the Fax Machine to the Modern Digital Age

The first discussions of an internal target experiment at HERA began
in 1987. In the 1980s, physicists typically used large mainframes (e.g.
VAX computers made by Digital Corporation, Maynard, MA). Each
physicist had a VT100 terminal at their desk. At that time, electronic
mail was in its infancy. Bitnet, a cooperative US university network
developed in the early 1980s, allowed brief exchange of messages
only among users connected to the mainframe. Any kind of serious

Table 7.2:   Summary of the leadership of the HERMES collaboration.

Year	Spokesperson	Deputy Sp.	Technical coordinator	Analysis coordinator
1988–1993	R. Milner & K. Rith	N/A	E. Steffens	M. Düren
1993–1994	K. Rith	M. Vetterli	E. Steffens	M. Düren
1994–1995	J. van den Brand	M. Vetterli	E. Steffens	M. Düren
1995–1997	R. Milner	A. Brüll	A. Simon	M. Düren
1997–1999	E. Kinney		A. Simon	N. Makins
1999–2001	E. Steffens	E. Aschenauer	J. Stewart	M. Vincter
2001–2003	D. Ryckbosch	E. Aschenauer	J. Stewart	U. Stoesslein
2003–2006	E. Aschenauer	P. Di Nezza	J. Stewart	N. Makins
2007	J. Stewart		N/A	N. Makins
2008	E. Kinney	A. Fantoni	N/A	N. Makins
2008–2011	K. Rith		N/A	G. Schnell M. Contalbrigo
2011–2019	G. Schnell		N/A	M. Contalbrigo A. Rostomyan C. Van Hulse L. Pappalardo

discussion required a real conversation and transatlantic phone calls were expensive then. Milner and Rith had several such calls in 1988 as the HERMES experiment was originated. International conferences were essential as a meeting place where in-person discussions could take place. For example, the Intersections conference at Rockport, Maine in May 1988 is credited as the birthplace of HERMES as it was the first time that the principals met in-person and could discuss the possibility of an internal target experiment at HERA. Presentations were made using colored pens and plastic transparencies displayed via an overhead projector.

At that time, exchange of large documents via email was not possible. Thus, the HERMES proposal was developed using the fax machine. Drafts were written using the word processor program TeX and the output was transmitted via fax. The document was marked up by pen and ink, and then sent back via fax. The final copy was sent via airmail and printed and bound locally. All figures were drawn in pen and ink.

By the mid-1990s, personal computers (typically Macs made by Apple) were popular and connected via telephone lines using a modem. The World Wide Web, developed in 1989 by Tim Berners-Lee's group at CERN, was available in 1994 on the workstations in the HERMES counting house at the HERA East Hall. The web pages used HyperText Markup Language (HTML) and the Mosaic web browser allowed communication across the web. The HERMES experiment had a web site in 1994. Further, as HERMES turned on, Microsoft PowerPoint became widely used as a program to format presentations. Initially, this was used to prepare plastic transparencies. However, by the late 1990s individual laptop personal computers became common and then PowerPoint Presentations could be displayed from the laptop via a digital projector.

In the mid-1990s, when the HERMES experiment turned on, cell phones were unknown. Pagers, wireless devices that can receive and display alphanumeric and voice messages, were commonly used. In particular, the HERMES leadership and those on the frontlines of running the experiment communicated via these pagers continuously. They were essential tools when unexpected events associated with the smooth operation of the experiment demanded immediate attention. Being Run Coordinator meant carrying a pager, which could ring in the middle of the night when an urgent issue arose. If it was not possible to fix it over the phone, the Run Coordinator had to make the trip to DESY.

In the mid 1990s, video conferencing became possible. However, the equipment was large and expensive, the quality was poor compared to modern standards and scheduling such a transatlantic call was not trivial. The call itself was also quite expensive.

Support of the HERMES collaboration was provided by a number of excellent members of the DESY administrative staff. Sabine Krohn has worked in the HERMES office in building 1e for over two decades and has welcomed and helped many new HERMES collaborators to settle into life at DESY. She has been ably assisted by Ramona Matthes and Soerne Moeller. In the 1990s, Suzanne Surrow worked in the HERMES office and later Phyllis Court also worked there.

7.4.1 *HERMES collaboration meetings*

Starting from the initial meeting at Rockport, Maine in May 1988, the HERMES collaboration met several times per year at locations on both sides of the Atlantic. Of course, the majority of meetings took place at DESY, particularly during the intensive years 1994–1996 when the experiment was being built in the East Hall. The DESY guest house was a convenient and inexpensive place to stay for visitors but was often full as the other HERA experiments were also active. The Hotel Schmidt in nearby Othmarschen was frequently used by HERMES collaborators and in the nearby Wagner bar there were many technical discussions of the HERMES design that continued late into the evening over some beers. HERMES collaborators could be frequently found eating the delicious Argentinian beef in the nearby Blockhouse. Around the corner on Waitzstrasse were a number of other good restaurants. In the early days, HERMES collaboration dinners were typically held in restaurants in proximity to DESY. The excellent seafood restaurants on Strandweg in Blankenese were often the location. The Ottensen quarter near Altona was another frequent location. The Groninger Pils brewery restaurant in Hamburg was another favorite where traditional and tasty German food could be washed down with beer from a keg brought to the table and then tapped.

In the early 1990s, HERMES collaboration meetings were held at the home institutes of members of the collaboration, e.g. at Heidelberg, Urbana-Champaign, and Cambridge, MA. In August 1990, the HERMES collaboration met in Vancouver, Canada and watched as the hostilities of the First Gulf War got underway. In June 1993, the HERMES collaboration met in Rome before the 13th International Conference on Particles and Nuclei which took place later in Perugia. Our Italian colleagues, under the direction of Salvatore Frullani, produced a sumptuous feast with ornately printed menus as we overlooked the sevens hills of Rome.

From January 1994 to January 1996, there were seven HERMES collaboration meetings at DESY. In 1994, the test experiment was installed and operated and later that year installation of the complete experiment got underway. The collaboration meeting in September 1994 was important as there it was decided to install a polarized ^{3}He target for the initial running of the experiment in 1995. The four meetings in January, May, July and November 1995 were intensive as the collaboration brought this completely new experiment into operation and started data taking. Interestingly, in the middle of the intensive commissioning of the experiment in May 1995, there were presentations on a possible RICH upgrade by Harold Jackson and report from the Laser Driven Source task force including Argonne, Erlangen, Illinois and TRIUMF.

By May 1996, the first year's data analysis was mature, and the polarized hydrogen target had been installed so the collaboration took itself to Pasadena, CA for its regular collaboration meeting. A focus of the meeting was the preparation of the publication on the 1995 data. There was a report on a visit to DESY by MIT theorist Bob Jaffe as well as one on the activities of the Laser Driven Source task force. Interestingly, there were presentations on studies of new physics accessible with a polarized internal gas target in the HERA proton ring as well as a study of physics with nuclear beams in HERA. At the Caltech meeting, there was a soccer game played between a team from HERMES, shown in Fig. 7.13 and a team from SLAC experiment E154, led by Emlyn Hughes, who was then on the faculty at Caltech. The final score was HERMES: 3 and E154: 2. Afterwards, the players from both teams shared a pleasant beer as we looked up at the San Gabriel mountains. Note that the two co-authors played on the winning HERMES team.

Regular collaboration meetings at DESY continued interspersed with meetings off-site. In February 1997, the HERMES collaboration met in Boulder, Colorado and in May 1997 met at DESY-Zeuthen. While driving through Berlin in May 1997, the visitor was met with a vast number of tall cranes as that city was in the midst of an enormous construction period. At Zeuthen, the HERMES soccer team played the local firemen in an enjoyable game.

Subsequently, HERMES collaboration meetings took place off-site at Amsterdam in November 1997, at Vancouver in February 1998, and at Heidelberg in May 1998, at Las Cruces, New Mexico

Fig. 7.13: The HERMES soccer team that played against the SLAC E154 team at Caltech in May 1996. The players were (left to right): back row: Bruce Bray, Kalen Martens, Richard Milner, Marc Beckmann, Mike Vetterli, Wolfgang Lorenzon, Eric Belz; front row: Ralf Kaiser, Johan Blouw, Greg Rakness, Michael Spengos, Armand Simon, Gunar Schnell, Erhard Steffens (*Source*: HERMES Coll.).

in February 1999, at Frascati in June 1999, at MIT in June 2000, at Ghent in August 2000, at Ferrara in May 2001, at Dubna in August 2011, at Yerevan in July 2002, at Bari in September 2002, at Warsaw in June 2004, at Erlangen in October 2005, and at Tokyo in September 2006.

In June 2007, when HERA ceased operation, the HERMES collaboration organized a scientific symposium and End of Data Taking (EDT) party at DESY. To the sound of the Doors "The End" playing through the HERA tunnel and East Hall counting room, the final HERMES data were written to disk. The two-day symposium had talks on the origins of the HERMES experiment, the personal reminiscences of the leaders over twelve years and scientific overviews of the HERMES results. There was an afternoon party barbecue in the parking lot of the East Hall on Saturday 30 June and a dinner at Das Bauernhaus in the Altona Volkspark

Fig. 7.14:   HERMES soccer players after the game at the end of data taking party in June 2007. Five members of the HERMES team that played at Caltech eleven years earlier are included (*Source*: HERMES Coll.).

on Sunday 1 July. Of course, there was a soccer game on the DESY soccer field and the players are shown in Fig. 7.14. Three annual HERMES collaboration meetings continued from 2007 to 2018 and the most recent collaboration meeting was held at DESY in December 2020. During this period, the HERMES collaboration met off-site at Beijing in October 2007, at Zeuthen in April 2011 and at Bilbao in October 2012.

7.4.2   HERMES offsite meetings

The younger HERMES people on-site in Hamburg regularly attended a social gathering, known as the *HERMES Offsite Meeting*, which typically took place in a nearby bar or restaurant on Thursday evenings. There was a dedicated website for the Offsite Meeting and the weekly social interaction provided an opportunity for the younger people to get to know each other personally. Many lifelong friendships were created here. For example, in June 2007, the Offsite Meetings took place at 21:00 at: Cafe Strand (June 14), at the Stadcafe Ottensen (June 21) and at the Titanic (June 28).

7.4.3   HERMES editorial board

An important standing committee, known as the HERMES Editorial Board, oversaw the preparation and publication of scientific results.

Fig. 7.15: Three long-time leaders of the HERMES experiment: Andy Miller from TRIUMF, Dirk Ryckbosch from Ghent, and Wolf-Dieter Nowak from DESY/Zeuthen, enjoying the end of data taking party in June 2007 (*Source*: HERMES Coll.).

The exacting standards imposed by this committee ensured that the HERMES publications obtained a justifiable reputation for high quality internationally. The review and approval by the Editorial Board was regarded by HERMES collaborators as the most challenging step in publishing scientific results.

Long-term members of the HERMES Editorial Board included Andy Miller, Gerard van der Steenhoven, Dirk Ryckbosch, and Wolf-Dieter Nowak (Fig. 7.15).

7.5 A New Generation of Physicists

The HERMES experiment provided particularly fertile ground for young physicists to establish their own careers in research and education in physics. An impressive list of more than 75 of such physicists with positions in laboratories and universities in Asia, Europe and N. America includes:

Moscow Amarian (Old Dominion University), Elke Aschenauer (Brookhaven), Harut Avakian (Jefferson Lab), Alessandro Bacchetta (U. Pavia), Christian Baumgarten (PSI), Andrea Belz

(Dvoredsky) (U. Southern California), Nicola Bianchi (INFN Frascati), Johan Blouw (MPI Heidelberg), Alexander Borissov (Pusan National U.), Antje Bruell (Jefferson Laboratory), Henk-Jan Bulten (NIKHEF), Marco Contalbrigo (INFN Ferrara), Wouter Deconinck (U. Manitoba), Markus Diefenthaler (Jefferson Lab), Pasquale Di Nezza (INFN Frascati), Michael Düren (U. Giessen), Alessandra Fantoni (INFN Frascati), Massimiliano Ferro-Luzzi (CERN), Horst Fischer (U. Freiburg), Brendan Fox (U. Hawaii at Manoa), Haiyan Gao (Duke U.), Erika Garutti (DESY), David Gaskell (Jefferson Lab), Michel Guidal (Orsay), Kawtar Hafidi (Argonne), Jens-Ole Hansen (Jefferson Lab), Delia Hasch (INFN Frascati), Holger Ihssen (Helmholtz Association), Sylvester Joosten (Argonne), Ralf Kaiser (U. Glasgow), Edward Kinney (U. Colorado-Boulder), Alexander Kisselev (Brookhaven), Uta Klein (U. Liverpool), Norbert Koch (TU Nürnberg), Hauke Kolster (Genentech), Wolfgang Korsch (U. Kentucky), Laird Kramer (Florida International U.), Inti Lehmann (GSI/FAIR), Paolo Lenisa (U. Ferrara), Wolfgang Lorenzon (U. Michigan), Allison Lung (Jefferson Lab), Naomi Makins (U. Illinois at Urbana-Champaign), Jeff Martin (U. Winnipeg), Kevin McIlhany (US Naval Academy), Richard Milner (MIT), Valeria Muccifora (INFN Frascati), Alexander Nagaitsev (Dubna), Eugenio Nappi (INFN Bari), Alexander Nass (Forschungszentrum, Juelich), Luciano Pappalardo (U. Ferrara), Vassili Papavassiliou (New Mexico State U.), Stephen Pate (New Mexico State U.), David Potterveld (Argonne), Mark Pitt (Virginia Tech U.), Greg Rakness (Fermilab), Davide Reggiani (PSI), Paul Reimer (Argonne), Caroline Riedl (U. Illinois at Urbana-Champaign), Patrizia Rossi (INFN Frascati and Jefferson Lab), Gunar Schnell (U. Bilbao), Marc Schumann (U. Freiburg), Ralf Seidl (RIKEN), Björn Seitz (U. Glasgow), Taeksu Shin (Institute for Basic Science, Korea), James Sowinski (DOE), Jörn Stenger (PTB Braunschweig), James Stewart (Brookhaven), Bryan Tipton (MIT Lincoln Labs), Anne Trudel (TRIUMF), Michael Tytgat (U. Gent), Johannes van den Brand (University of Maastricht), Gerard van der Steenhoven (U. Twente), Charlotte Van Hulse (CERN), Michel Vetterli (Simon

Fraser U.), Manuella Vincter (Carleton U.), Jan Visser (NIKHEF), Thomas Wise (Madison), Kirsten Zapfe (DESY), Beni Zihlman (Jefferson Lab).

Many more have had productive and fulfilling careers across the full range of human activity and interest.

7.6 Epilogue

HERMES was a technically innovative experiment created by a transatlantic collaboration to study the fundamental spin structure of matter. It's aim to approach the ideal high-energy electron scattering experiment with full control of the beam and target spins could only be realized at the HERA collider at DESY, Hamburg, Germany. The development of highly-polarized (both transverse and longitudinal), sufficiently thick internal gas targets was an essential aspect as well as the highly self-polarized electron beam in HERA. The implementation of hadron identification and particle identification allowed the comprehensive measurement of semi-inclusive scattering for the first time (Fig. 7.16).

In parallel with HERMES becoming a reality, theorists worked out the implications of semi-inclusive measurements and developed the realization that the transverse structure of the nucleon could be accessed. New parton distributions and momentum distributions were defined and HERMES data both constrained and stimulated further theoretical work. Other experiments were quick to join in and what resulted was a transformation in how we think about proton structure. This has shaped current research and has helped to motivate the proposal to construct a new US-based accelerator to study the fundamental structure of matter well into the twenty first century.

Writing this book has been a labor of love. We have learned that, among many of the people who have worked on HERMES, there is an emotional attachment to the experiment, the time, the place and the colleagues who carried it out. As you have read, many who started their careers at HERMES have advanced to be leaders in academia,

Fig. 7.16:   Klaus Rith, Richard Milner and Erhard Steffens on January 16, 2020 at the celebration of the 60th birthday of the DESY laboratory at Hamburg City Hall (*Source*: R. Milner).

research and industry. However, for most in the HERMES family, there is a strong feeling that the memories of the time spent working together on that experiment in the HERA East Hall retain a special magic. In this book, we have tried to capture some of the magic of HERMES.

Appendix A

Primer on Units, Scientific Notation, and Technical Terms

A.1 International System of Units

The SI (from the French *Système International d'Unités*) system of units is founded on the six basic quantities in Table A.1. The system allows for an unlimited number of units based on the basic quantities.

Many units are called after famous physicists: e.g. Joule (unit of energy), Coulomb (unit of charge), Ampere (unit of current), and Newton (unit of force).

Since 2019, the magnitudes of all SI units have been defined by declaring exact numerical values for seven defining, fundamental constants, when expressed in terms of their SI units. These include:

- the speed of light in vacuum $c = 299\ 792\ 458$ m/s,
- the Planck constant $h = 6.626\ 070\ 15 \times 10^{-34}$ J·s,
- the elementary charge on the electron $e = 1.602\ 176\ 634 \times 10^{-19}$ C,
- the Boltzmann constant $k = 1.380\ 649 \times 10^{-23}$ J/K,
- the Avogadro constant $N_A = 6.022\ 140\ 76 \times 10^{23}$ mol^{-1}.

As we will see below, the speed of light is essential to determining the energies of subatomic particles. The Planck constant enters whenever quantum mechanics must be used to describe physical phenomena. The elementary charge on the electron (and proton) is a universal constant. The Boltzmann constant relates the temperature T of a large number of atoms or molecules to the average energy. Finally, a mole contains N_A molecules or atoms.

Table A.1:   SI base units.

Quantity	Name	Symbol
length	meter	m
mass	kilogram	kg
time	second	s
electric current	ampere	A
temperature	kelvin	K
amount of substance	mole	mol

It is important to understand the concept of energy as it applies to different physical situations. The energy due to the speed of a macroscopic object like a runner or a car is called *kinetic energy* and is typically given in *Joules*, denoted by the symbol J, where

$$1 \, \mathrm{J} = 1 \, \mathrm{kg} \, \mathrm{m}^2/\mathrm{s}^2.$$

A runner of mass $70 \, \mathrm{kg}$ and moving at a speed of $2 \, \mathrm{m \, s^{-1}}$ (typical jogging speed) has a kinetic energy of $140 \, \mathrm{J}$. Now Einstein showed that any object of mass M has an associated *rest energy* Mc^2, where c is the speed of light given above. This rest energy becomes important if it is comparable to the kinetic energy. The runner has a rest energy of

$$70 \cdot (3 \times 10^8)^2 = 6.3 \times 10^{18} \, \mathrm{J},$$

which is huge compared to the runner's kinetic energy. The runner's speed is tiny compared to the speed of light so relativistic effects are truly negligible.

Consider now subatomic particles. A charge q when accelerated across a potential V acquires a kinetic energy of qV. For example, an electron (or positron) of charge e, given above, when accelerated through $10^6 \, \mathrm{V}$ ($1 \, \mathrm{MV}$) acquires a kinetic energy of $1 \, \mathrm{MeV}$, where

$$1 \, \mathrm{MeV} = 10^6 \cdot 1.6 \times 10^{-19} = 1.6 \times 10^{-13} \, \mathrm{J},$$

which is tiny compared to the kinetic energy of the jogger. However, now consider the rest energy of the electron (or positron). The mass is 9.1×10^{-31} kg giving a rest energy of 0.511 MeV, which is comparable to 1 MeV kinetic energy. From Einstein's Theory of Special Relativity, the speed of the object becomes *relativistic*, i.e. close to the speed of light. For an electron (or positron) with 1 MeV kinetic energy, its speed is 94% of c.

The energy quantity eV (pronounced as electron-Volt) is widely used in subatomic physics. 1 eV is the kinetic energy acquired by an electron when accelerated through 1 V. For example, the electron and proton in the hydrogen atom are bound by 13.6 eV. This is the energy scale of electrons in atoms and molecules in chemical reactions. Burning 1 kg of coal releases about 30 MJ, which corresponds to about 4 eV per atom burned. Nuclear energies are much higher. The energy released in the fission of Uranium-235 is about 200 MeV, or about 50 million times larger than the energy released in the burning of a carbon atom. The chemical energy reactions drive life on Earth while the nuclear energy processes fuel the stars like the Sun.

Consider now the proton. Its mass is 1.7×10^{-27} kg or 938 MeV/c^2, which is 1836 times the mass of the electron. A proton with 1 MeV kinetic energy has a speed which is 4% of c which is very non-relativistic. A proton of 1 GeV kinetic energy has a speed of 87% of c.

The HERA accelerator stored electron and positron beams of 27 GeV kinetic energy and proton beams of energy 920 GeV. Both are highly relativistic, i.e. have speeds very close to c.

A.2 Scientific Notation

Scientific notation is a convenient way of writing very large or very small numbers. For example, in describing the length scales in the universe, physicists have to deal with distances from 8.8×10^{26} m, the approximate diameter of the visible universe, to 2×10^{-20} m, the shortest distance currently probed by physicists at the Large Hadron Collider. This is over 46 orders of magnitude! Table A.2 summarizes commonly used notation used by physicists and gives examples of how they are used.

Table A.2: The 20 SI prefixes used to form decimal multiples and submultiples of SI units.

Prefix	Symbol	Factor	Example
yocto	y	10^{-24}	1.7 yg is the approximate mass of the proton.
zepto	z	10^{-21}	Distances down to 20 zm are probed at the LHC.
atto	a	10^{-18}	0.35 asec is the time it takes light to travel across the hydrogen atom.
femto	f	10^{-15}	The proton radius is about 1 femtometer.
pico	p	10^{-12}	The radius of the hydrogen atom is 53 pm.
nano	n	10^{-9}	The time between electron bunches in HERA was 100 ns.
micro	μ	10^{-6}	The diameter of the human hair is about 100 μm.
milli	m	10^{-3}	A U.S. ten cent (dime) coin is about 1 mm thick.
kilo	k	10^{3}	The distance from Boston to New York by road is 355 km.
mega	M	10^{6}	The radioactive potassium in a banana decays by emitting a γ−ray of energy about 1 MeV.
giga	G	10^{9}	The diameter of the Sun is about 1 Gm.
tera	T	10^{12}	The proton beams at the LHC each have an energy of 7 TeV.
peta	P	10^{15}	10 Pm is the distance light travels in one year.
exa	E	10^{18}	0.4 Esec is the approximate age of the universe.
zetta	Z	10^{21}	1 Zm is the approx. diameter of the Milky Way galaxy.
yotta	Y	10^{24}	The size of the universe is 880 Ym.

A.3 Structure in the Universe

An interesting application of scientific notation is to quantify the length scales in the universe from the microcosm (smallest) to the cosmos (largest). The factor 10^5 allows one to scale approximately to different layers of structure. Starting from the human scale of 1 m, one scales down by 10^5 to the thinnest macroscopic material, e.g. household aluminum foil. A further reduction of 10^5 brings us to the typical size of an atom. Finally, a further reduction of 10^5 brings us to the proton.

Starting from the Earth's radius, and scaling up successively by 10^5 gives approximate length scales for the Earth-Sun distance, the

Table A.3:   Approximate length scales in the universe.

Object	Size m	Abbreviated notation
Proton size	10^{-15}	1 fm
Size of an atom	10^{-10}	100 pm
Household aluminum foil thickness	10^{-5}	10 μm
Height of a human	1	1 m
Radius of Earth	6.4×10^6	
Earth–Sun distance	1.5×10^{11}	1 AU
Size of Solar System	10^{16}	1 light year
Size of Milky Way Galaxy	10^{21}	10^5 light years
Radius of Universe	4×10^{26}	10^{10} light years

size of the Solar System, the size of our Milky Way galaxy, and the size of the universe, respectively.

For some unknown reason, matter is structured at length scales over 40 orders of magnitude with some regularity.

An Astronomical Unit (AU) is the approximate distance between the Earth and the Sun (93 million miles). Since 2012, it has been defined exactly as

$$1 \, \text{AU} = 149\,597\,870\,700 \, \text{m}.$$

A light year is the distance that light travels in 1 year and equals 9.5×10^{15} m.

A.4   Explanation of Commonly-Used Technical Terms

Here, we provide a brief explanation of the more commonly used technical terms in the book. The reader should be familiar with the material in Chapter 1 and should refer to Fig. 1.4, where the basic elements of the Standard Model are summarized.

Asymmetry: In scattering of polarized electrons from polarized nucleons, the scattering rate depends on the polarizations of both beam and target. The asymmetry is the ratio of the difference in scattering rates divided by the sum.

Baryon: A baryon is a composite subatomic particle which contains an odd number of valence quarks, e.g. three, as in the case of the proton and neutron.

Boson: A boson is a quantum mechanical particle with integer spin, i.e. $0, 1, 2, \ldots$ The force carriers of the Standard Model are all bosons. Other examples include: the photon, the deuteron, the Higgs boson and the helium-4 nucleus.

Cross section: The cross section is a measure of the probability that a specific process will take place in the collision of two particles. It is expressed in terms of the transverse area that the incident particle must hit in order for the given process to occur. The Standard Model prescribes detailed procedures on how to calculate cross sections for quantum mechanical scattering of relativistic particles.

Fermion: A fermion is a quantum mechanical particle with half-integer spin, e.g. $\frac{1}{2}, \frac{3}{2}, \frac{5}{2}, \ldots$ The quarks and leptons of the Standard Model are all fermions. Other examples include: the proton and neutron, and the helium-3 nucleus.

Gluon: The gluon is the electrically uncharged, massless boson that mediates the strong force in Quantum Chromodynamics. It is unobservable directly.

Helicity: This is the projection of the spin on to the direction of momentum of a subatomic particle. The helicity of a particle is right-handed if the direction of its spin is the same as the direction of its motion and left-handed if opposite.

Hadron: A hadron is a composite subatomic particle made of two or more quarks bound by the strong force. Both baryons and mesons are hadrons.

Kaon: A kaon is a meson composed of a *strange* quark and an antiquark.

Luminosity: Luminosity quantifies the number of interactions per second and is the proportionality factor between the number of scattering events per second and the cross section. In a fixed target experiment like HERMES, the luminosity depends on the beam current and the target areal density.

Meson: A meson is a hadron composed of a quark and antiquark bound by the strong force.

Muon: The muon is the heavier (207 times the electron mass) partner of the electron in the Standard Model. It is unstable with a lifetime of 2.2 microseconds, at rest.

Neutron: The neutron is a baryon with about the same mass as the proton but without an electric charge and is present in all atomic nuclei except for the hydrogen atom. It is composed of three valence quarks: two *down* quarks, and an *up* quark. In addition, it contains quark-antiquark pairs and gluons. The free neutron is unstable with a lifetime of about 15 minutes.

Nucleon: A nucleon is a proton or a neutron.

Photon: The photon is the massless exchange boson of the quantum theory of electricity and magnetism.

Pion: The pion is the lightest meson with a mass about 270 times that of an electron. It is denoted by the Greek symbol π and comes in three charge states: π^+, π^0, and π^-. Charged pions decay with a mean lifetime of about 26 nanoseconds (at rest) into muons and muon neutrinos. The neutral pion decays into two photons with a much shorter lifetime of 84 attoseconds (at rest).

Polarization: Spin polarization is the degree to which the spin, of the beam particles or the target atoms, is aligned with a given direction.

Proton: The proton is the nucleus of the hydrogen atom with a positive electric charge equal in magnitude to that of the electron but opposite in sign. It is a baryon composed of three valence quarks: two *up* quarks, and a *down* quark. In addition, it contains quark–anti-quark pairs and gluons. The proton is stable.

Quark: The quarks are the elementary, structureless particles of the Standard Model that combine to form hadrons. Quarks have various intrinsic properties, including electric charge, mass, color charge, and spin. They are not found as free particles. They are the only elementary particles in the Standard Model to experience all four fundamental interactions, as well as the only known particles whose electric charges are not integer multiples of the elementary charge.

Sea quarks: In a hadron, sea quarks arise from the virtual quark–anti-quark pairs that are produced in the internal interactions.

Sub-atomic: The sub-atomic scale is the domain of physical size that encompasses objects smaller than an atom. It is the scale at which the atomic constituents, such as the nucleus containing protons and neutrons, and the electrons, which orbit in paths described by quantum mechanics around the nucleus, become apparent.

Valence quarks: These are the quarks and anti-quarks that give rise to the quantum numbers of the hadron.

Appendix B

List of Acronyms

ABS: **A**tomic **B**eam **S**ource
The sophisticated device that produces a beam of spin polarized atoms.

BNL: **B**rookhaven **N**ational **L**aboratory
The nuclear physics, accelerator and inter-disciplinary laboratory on Long Island, New York, USA.

CEBAF: **C**ontinuous **E**lectron **B**eam **F**acility
The high duty-factor electron accelerator using superconducting radiofrequency cavities at Newport News, Virginia, USA.

CERN: **C**onseil **E**uropéen pour la **R**echerche **N**ucléaire
The high-energy accelerator and particle physics laboratory in Geneva, Switzerland.

DESY: **D**eutsches **E**lektronen **SY**nchrotron
The electron accelerator and photon science laboratory in Hamburg, Germany.

DIS: **D**eep **I**nelastic **S**cattering
The high-energy lepton scattering process where the charged constituents of the proton, neutron and nuclei are directly probed.

DVCS: **D**eeply **V**irtual **C**ompton **S**cattering
The high-energy lepton scattering process on the proton where a high-energy photon is produced in addition to the scattered electron and the proton is left intact.

EIC: Electron-**I**on **C**ollider
The high-luninosity, polarized electron-polarized ion collider planned for construction at Brookhaven National Laboratory, New York, USA.

EMC: European **M**uon **C**ollaboration
European collaboration that carried out ground-breaking experiments on the fundamental structure of nuclei and the origin of proton spin at CERN in the 1970s and 1980s.

GPD: Generalized **P**arton **D**istribution
Generalization of the partons of Bjorken and Feynman to include the effects of spin and transverse momentum.

HERA: Hadron **E**lektron **R**ing **A**nlage
World's only electron/positron-proton collider operated at DESY from 1992 to 2007.

HERMES: HERA **ME**asurements with **S**pin
Technically innovative experiment to study proton spin structure at East Hall of the HERA collider at DESY that took data from 1995 to 2007.

ISPC: International **S**pin **P**hysics **C**ommittee
The Committee that coordinates the worldwide organization of major conferences and workshops on spin physics.

JLab: Jefferson **Lab**oratory
World's first large-scale superconducting accelerator and frontier electron scattering laboratory at Newport News, VA, USA that began operation in 1995 and was recently upgraded to higher beam energy.

LHC: Large **H**adron **C**ollider
World's high-energy frontier particle physics accelerator that began operation in 2010.

LNS: Laboratory for **N**uclear **S**cience
MIT laboratory in which Physics Department faculty and their groups carry out research into nuclear and particle physics.

MPI-K: Max-**P**lanck **I**nstitute für **K**ernphysik-Heidelberg
German institute that played a key role in the origin of HERMES.

MEOP: Metastability **E**xchange **O**ptical **P**umping
Technique invented in Texas, USA in 1963 to transfer angular momentum from a light beam to polarize ^{3}He nuclei.

PGT: Polarized **G**as **T**arget
Localized, high-density configuration of neutral atoms with spin-polarized nuclei arranged to allow scattering of an incident intense charged particle beam.

PRC: Physics **R**esearch **C**ommittee
The international advisory and review committee reporting to the DESY Directorate.

PSI: Paul-**S**cherrer **I**nstitute for basic and applied research with nuclear methods, Zürich, Switzerland.

PSTP: Workshops on **P**olarized **B**eams, **T**argets and **P**olarimetry.

RHIC: Relativistic **H**eavy **I**on **C**ollider
World's first heavy-ion collider and only polarized proton collider at Brookhaven National Laboratory, Long Island, New York, USA, that began operation in 2000 and is planned to be the basis of the U.S.-based EIC.

RICH: Ring **I**maging **CH**erenkov
Particle detector that measures photons emitted as rings concentric about the path to determine the particle's velocity.

SIDIS: Semi-**I**nclusive **D**eep **I**nelastic **S**cattering
High-energy electron scattering DIS process from a proton where a recoil hadron is detected in coincidence with the scattered electron.

SLAC: Stanford **L**inear **A**ccelerator **C**enter
Electron accelerator and photon science laboratory in Menlo Park, California where DIS was first discovered by the MIT-SLAC collaboration in 1967.

TMD: **T**ransverse **M**omentum **D**istribution
Distribution of quarks and gluons in the proton in the plane
transverse to the direction of the exchanged virtual photon.

TSR: **T**est **S**torage **R**ing for the study of laser cooling of ions in
different charge states, employed for proton spin-filtering studies in
1992; MPI-K, Heidelberg, Germany.

Appendix C

A Short History of Spin

This appendix outlines the historical development of spin in physics from about 1920 to the present day. It aims to provide the reader with an accurate chronology of important developments, both scientific and technical. It was written by Richard Milner as a contribution to the proceedings of the XVth International Workshop on Polarized Sources, Targets, and Polarimetry, held on 9–13 September 2013 at the University of Virginia, Charlottesville, Virginia, USA.

C.1 Introduction

Classical mechanics, specifically rigid body motion, contains the ideas of spin and its relationship to angular momentum. For example, in astronomy spin-orbit coupling between celestial objects reflects the conservation law of angular momentum. Around 1920, the study of atomic systems led to a failure of classical mechanics and the emergence of an entire new paradigm to describe the subatomic world, namely quantum mechanics. Spin was found to be an essential quantum number of all subatomic particles and its profound properties underpin the physicist's present-day understanding of the structure of matter. In this paper, I provide a brief outline of the history of spin from 1925 to the present day both in terms of the scientific understanding of the fundamental structure of matter as well as the development of key experimental tools. I have found it useful to relate the history as a narrative involving four consecutive time periods.

C.2   Explaining the Fundamental Structure
of Matter: 1925–1950

In the early 20th century it became evident that atoms and molecules with even numbers of electrons are more chemically stable than those with odd numbers of electrons. For example, Lewis in the third of his six postulates of chemical behavior stated[20] that the atom tends to hold an even number of electrons in the shell which surrounds the nucleus. In 1919, Langmuir suggested[21] that the periodic table could be explained if the electrons in an atom were connected or clustered in some manner.

About 1920, the Bohr-orbit theory provided the accepted understanding of the atom by physicists. However, the number of states observed experimentally was double what was predicted by Bohr–Sommerfeld quantization rules. This mysterious doubling was known as *Mechanische Zweidentigkeit* in German and as *duplexity* in English. In 1922, Stern and Gerlach carried out their famous experiment[22] of passing silver atoms through an inhomogeneous magnetic field and observing a deflection either up or down by the same amount. At the time, the experiment was interpreted as a crucial validation of the Bohr-Sommerfeld theory over the classical theory of the atom. It showed clearly that spatial quantization exists, a phenomenon that can be accommodated only within a quantum mechanical theory. In 1924, Bose, at Dhaka University, derived Planck's quantum radiation law by counting states with identical properties and sent it to Einstein, who translated it into German himself, and had it published in Zeitschrift für Physik.[23] In 1925, the Pauli Exclusion Principle was formulated[24] as: *no two electrons can have identical quantum numbers.*

Most significantly for this discussion, also in that year, Leiden graduate students Uhlenbeck and Goudsmit first hypothesized[25] intrinsic spin as a property of the electron. This occurred over the strong objections of some prominent physicists but with the support of their advisor, Paul Ehrenfest. In their Nature letter they write: *It seems that the introduction of the concept of the spinning electron makes it possible throughout to maintain the principle of the successive building up of atoms utilized by Bohr in his general discussion of the relations between spectra, and the natural system of the elements. Above all, it may be possible to account for the important results arrived at by Pauli without having to assume an unmechanical duality in the binding of the electrons.* In the succeeding letter in the same journal, Bohr fully agreed.

In 1926, Thomas[26] correctly applied relativistic calculations to spin–orbit coupling in atomic systems and resolved a missing factor of two in the derived g-values. Also in 1926, Fermi[27] and Dirac[28] developed the Fermi–Dirac statistics for electrons. It was immediately applied to describe stellar collapse to a white dwarf,[29] to electrons in metals,[30] and to field electron emission from metals.[31]

In 1928, Dirac developed[32] his elegant equation for spin-$\frac{1}{2}$ particles. In this formulation, solutions are four-component spinors which are interpreted as positive and negative energy states of spin $\pm\frac{1}{2}$ each. Dirac predicted the existence of the positron, and the theory became the basis for the most precisely tested theory in physics, Quantum Electrodynamics. By the end of the 1920s, physicists had developed a fundamental understanding of the essential role of electron spin in explaining the electronic structure of the atom. There exist excellent, personal, historical accounts by Dirac,[2] Uhlenbeck,[3] and Goudsmit[4] of this period.

In 1927, Wrede, a student of Stern at Hamburg,[33] and Phipps and Taylor at Illinois[34] independently observed the deflection of atomic hydrogen in a magnetic field gradient. In 1929, Mott wondered[35] if electron spin can be observed directly via the scattering of electrons from atomic nuclei. Note that in the appendix to his paper, Mott showed that the Stern-Gerlach experiment cannot be carried out for electrons. Only in 1942 did Shull *et al.* verify[36] Mott's prediction in a double scattering experiment which used 400 keV electrons from a Van de Graaf generator. In the mid-1920s, Heisenberg and Hund postulated the existence of two kinds of molecular hydrogen: *orthohydrogen* where the two proton spins are aligned parallel and *parahydrogen* where the two proton spins are antiparallel. By the end of the decade, they had been studied experimentally. Later, by deflection of orthohydrogen in a magnetic field gradient, Stern and collaborators measured the g-factor of the proton to be about 2.5 nuclear magnetons,[37] a marked deviation from the Dirac value for a pointlike spin-$\frac{1}{2}$ particle, and the first hint of its internal structure.

In the 1930s, Rabi and collaborators (inc. N. Ramsey and J. Zacharias) using molecular beams in a weak magnetic field measured the magnetic moments and nuclear spins of hydrogen, deuterium, and heavier nuclei.[38]

By the end of the 1940s, the nuclear shell model had been established.[39] This explained the properties and structure of atomic nuclei and underscored the essential role of proton and neutron spin. A key aspect was the strong role of spin–orbit coupling, which was suggested to Goeppert–Mayer by a question from Fermi.

C.3   Developing Spin as an Experimental Tool: 1950–1975

By the middle of the twentieth century, the intrinsic spin of subatomic particles was a cornerstone of the physicist's theoretical understanding of the fundamental structure of matter. However, spin as an experimental tool became a reality only in this second era. In the 1950s, a number of seminal experiments were carried out using spin. In 1956, Lee and Yang pointed out that parity should be violated in the weak interaction.[40] Shortly afterwards, in 1956, Wu and collaborators observed[41] parity violation in aligned ^{56}Co. In 1958, it was shown experimentally[42] using polarization techniques that the neutrino has negative helicity. In 1959, the Thomas–Bargmann–Michel–Telegdi equation describing the spin precession of an electron in an external electromagnetic field was derived.[43]

In the 1960s, the discovery of point-like constituents in the proton at SLAC using deep inelastic scattering (DIS) profoundly affected our understanding of the fundamental structure of matter. A key determination that these constituents had spin-$\frac{1}{2}$ led to their identification as the quarks of SU(3) symmetry. Important sum rules related to spin-dependent DIS were derived by Bjorken[44] and by Ellis and Jaffe.[45]

During this period, the international spin community grew significantly in size to become the active, subfield of international physics we have today. Beginning in 1960 at Basel, symposia on polarization phenomena in nuclear reactions were held every 5 years until 1994. Beginning in 1974 at Argonne, symposia on high energy spin physics were held every 2 years until 1994. Beginning in 1996 in Amsterdam, the international spin community became unified and a symposium on spin physics has been held every 2 years since then. The International Spin Physics Committee was formed to oversee the organization of this biennial symposium which was held most recently in Ferrara, Italy in 2018. The next meeting is scheduled to take place in Matsue, Japan in 2021. The published proceedings of these meetings form the essential record of the research activities over this time. In Ref. 46, there is a complete tabulation of these meetings as well as references to their proceedings. Further, important conventions at Basel in 1960 for spin-$\frac{1}{2}$ particles[47] and at Madison in 1970 for spin-1 particles[48] were established to facilitate consistent discussion of spin observables. In this period of 25 years, enormous progress was made in developing spin as an experimental tool. Highlights are summarized under a number of important developments as follows:

C.3.1 *Polarized ion sources*

This brief summary is based on the historical reviews by Haeberli in 1967[49] and in 2007.[50] In 1946, Schwinger suggested[51] using double scattering of neutrons to determine the sign of the spin–orbit coupling. Wolfenstein pointed out[52] that using protons was more practical. The first experiment was carried out by Heusinkveld and Freier[53] to resolve the Fermi-Yang ambiguity,[54] which arises in the analysis of the scattering between particles of spin-$\frac{1}{2}$ and spin 0.

The atomic beam source (ABS) to produce polarized atoms became a reality in this period. In 1951, Paul proposed to use magnetic multipoles to focus atomic beams. Experimental work directed at the preparation of polarized-ion beams for nuclear experiments by the atomic-beam method was first undertaken by Clausnitzer,

Fleischmann, and Schopper in 1956.[55] Radiofrequency transitions were first developed at Saclay where the atomic beam method was to involve ionization in the cyclotron magnetic field. Generally, they utilized the *adiabatic fast passage method*, as proposed by Abragam and Winter.[67] By use of RF transitions, one can freely choose between vector and tensor deuteron polarizations. The development of the atomic beam source got underway at Erlangen in 1958 by Fleischmann's group. Further, in 1964, Gruebler, Schwandt, and Haeberli developed[57] the first source of polarized H$^-$. The ABS produced an intensity of $\approx 10^{16}$/s with high polarization ($\approx 90\%$) by 1970. In addition, polarized proton sources using the hydrogen metasable state were invented in the 1960s.[49]

Further, deuteron sources played an important role in spin physics. The first operational polarized ion source was the Basel deuteron source which operated in 1960 already. Measurement of the first asymmetries was carried out using a polarized beam from an ABS. In contrast to the Erlangen approach, Huber's group at Basel reduced the background signal by using deuterons. In addition, in 1975, there was already a polarized ^{6}Li beam at the Heidelberg tandem accelerator which was used for scattering experiments. Later, ^{7}Li and ^{23}Na beams were used at tandems to measure reactions including sub-barrier fusion of aligned ^{23}Na ions.

C.3.2 *Polarized electron sources*

The brief summary here is based on the historical review of the development of polarized electrons sources by Prescott.[58] In 1963, Hughes and colleagues begin consideration of polarized electron sources at Yale.[59] The initial work was based on production of polarized electrons by photoionization of a polarized atomic beam of alkali atoms. This effort produced the PEGGY source based on photoionization of ^{6}Li which was commissioned at SLAC in 1974. Sokolov-Ternov self-polarization was first observed in 1968 at the ACO storage ring in Orsay, France and shortly after that at Novosibirsk, USSR and Frascati, Italy. Further, a source of polarized electrons was developed[61] using optically oriented metastable atoms in a flowing helium afterglow.

C.3.3 *Polarized proton targets*

The brief summary here is based on the 2013 review of Keith.[62] While the earliest examples of polarized targets were polarized by static methods (i.e. low temperatures and high magnetic fields, see for example[63]), dynamic nuclear polarization (DNP) has been used for the majority of solid polarized targets used in nuclear and high energy physics. This technique began in 1953, when Overhauser at Illinois proposed to transfer the polarization of conduction electrons in a metal to nuclei by saturating the electron's spin resonance with RF radiation.[64] Initially met with great skepticism, Overhauser's suggestion was experimentally verified by Carver and Slichter later that year.[65] Working independently, Jefferies[66] and Abragam[68] both suggested to dynamically polarize nuclei by saturating so-called "forbidden transitions" in which electron and nuclear spins flip simultaneously. In 1962, Abragam, Borghini and co-workers built the first polarized proton target for the 20 MeV polarized proton beam at Saclay.[68] Shortly thereafter, Chamberlain, Jefferies, and collaborators built a polarized proton target for 250 MeV pion scattering experiments at Berkeley.[69] In both cases the average polarization of protons in the LMN target material was only about 20%, but steady improvement would be made in the following years, thanks to refinements in magnets and cryogenics.[70]

In the late 1960s, efforts were made to develop target materials with a higher percentage of polarized free protons (compared to the 3% in LMN). This work was led in part by Borghini at CERN and motivated by advances in the spin-temperature description of the DNP process.[71] It culminated in 1974, when de Boer and Niinikoski demonstrated 98% proton polarization in propanediol doped with the Cr(V) free radical.[72] By the early 1970's, diols and alcohols had replaced LMN as the target material of choice, due to their higher free proton content, higher resistance to radiation damage, and higher polarizations. Around the same time, the first highly successful frozen spin target was built at CERN by Niinikoski and Udo[73] and utilized advances that Niinikoski had made in ^3He–^4He dilution refrigeration.[74] In 1979, Niinikoski and Rieubland produced proton polarizations greater than 90% in irradiated NH_3,[75] which would soon become a common target material (especially for higher luminosity experiments), thanks to its even higher proton content and radiation resistance.[76] Further details about the current status of polarized proton targets can be found in the comprehensive review by Crabb and Meyer.[77]

Note that later development work, see Section C.4.3, produced highly polarized hydrogen, deuterium, and ^{3}He gas targets for use with stored beams in storage rings. Finally, it must be noted that in the early 1970s, proton spin was proposed as a diagnostic tool for medicine. This evolved into the technique of Magnetic Resonance Imaging, now in use daily worldwide.

C.3.4   *Polarized ^{3}He beams and targets*

Important spin capabilities using spin-1 photons were invented in this period. In 1950, Kastler proposed[78] the technique of optical pumping and, in 1960, the laser was developed. The technique to polarize ^{3}He gas using metastability exchange optical pumping was developed by Colegrove, Shearer, and Walters in 1963.[79] Polarized ^{3}He sources and targets based on flash lamps were used in nuclear physics experiments. For example, the spin-dependence of the fusion cross section ^{2}H+^{3}He $\rightarrow$ ^{4}He+^{1}H was measured at Basel in 1971.[80] Magnetometers using polarized ^{3}He were developed and located beneath the oceans to detect submarines.

C.4   Using the Tool: 1975–1995

In this period, the tools invented in the previous era started to be employed to great effect. In particular, experiments using polarized beams at high energies at ANL, CERN, Fermilab and SLAC produced results which continue to shape our understanding of the fundamental structure of matter. At SLAC, the development of polarized electron sources initiated a program of measurements which continued until the end of the century. At CERN, the highly polarized, high energy muon beams resulting from inflight pion decay have resulted in a series of experiments EMC, SMC, and

COMPASS which have used DNP targets and have profoundly impacted our understanding of the structure of the nucleon. In addition at lower energies, polarized beams played a central rule in experimental studies, e.g. spectroscopy (resolving ambiguities in angular momentum assignments), determination of spin-dependence in the optical model, constraining the effects of the D-state in light nuclei,[81] and angular momentum dependence in stripping and pickup reactions.

C.4.1 *High energy lepton scattering*

The 1970s saw the SLAC experiments E80 and E130 measure spin-dependent inclusive DIS from the proton for the first time. It was found that the valence quarks in the proton were polarized as expected.[82] At SLAC plans to construct the SLC were underway, so a subsequent proposal led by Hughes to probe the valence quark region in the neutron was turned down. Hughes turned his attention to CERN where the EMC experiment was being mounted. It took data for about a decade and in 1988 produced spin-dependent DIS data on the proton at low x for the first time. The EMC data showed[83] that the quarks carried far less of the proton's spin than was expected and that the Ellis-Jaffe Sum Rule was violated. Thus, the "proton spin crisis" was born.

Also at SLAC, experiment E122 for the first time used a polarized electron source based on optical pumping of GaAs. This technology dramatically enhanced the ability to carry out experiments using polarized electron beams. At that time, the electron beam polarization was limited to about 40%. In 1978, E122 announced the observation of parity violating electron scattering at SLAC for the first time.[84] This validated the Weinberg-Salam Model and provided the first measurement of the neutral current coupling of the electron.

Following the discovery of our incomplete understanding of the proton's spin, four major campaigns at CERN/SMC/COMPASS, SLAC/End Station A, DESY/HERMES, and the future RHIC-spin were conceived in this period. In particular, at SLAC, the use of strained GaAs resulted in significantly higher polarization.[85]

Subsequently, this technology has been employed at MIT-Bates, Jefferson Lab, Mainz, and Bonn.

Finally, at SLAC an important series of measurements with polarized electrons at the Z-pole got underway. The use of polarization at SLAC allowed the SLC to compete with the significantly higher luminosity LEP experiments in carrying out precision tests of the Standard Model. At LEP, transverse polarizations via Sokolov–Ternov self-polarization in excess of 50% were observed and the effects of the Earth's tides on the beam tune were observed.[86]

C.4.2 *High energy proton scattering*

In this period, high energy proton–proton spin experiments were carried out at the ZGS, AGS, and Fermilab. This brief summary is taken from the review article by Krisch.[87] High energy polarized proton beams became available in 1974 at the Argonne ZGS.[88] Krisch and collaborators measured unexpectedly large asymmetries at high $p_\perp^2$ in pp elastic scattering in contradiction to the expectations from QCD.[89] To this day, these results do not have an accepted explanation. With the phaseout of the ZGS in the late 1970s, polarized protons were developed at the AGS at Brookhaven National Laboratory. It was far more difficult to accelerate protons in the strong-focusing AGS than in the weak-focusing ZGS. To accelerate polarized protons to $22\,\text{GeV}$ at the AGS, 45 strong depolarizing resonances were successfully overcome.

At Fermilab beginning in the 1970s, polarization experiments using protons were carried out. Inclusive hyperon polarization experiments showed[90] increasing polarization with p_T. These data are consistent with other experiments carried out at KEK and the CERN ISR. Further experiments at Fermilab carried out by Yokasawa and colleagues measured[91] A_N, the transverse single spin asymmetry, at $200\,\text{GeV}$ for inclusive forward $\pi^\pm$ meson production and measured increasing asymmetries vs. x_F. A_N measurements vs. x_F at $\sqrt{s} = 4.9\,\text{GeV}$ at ANL,[92] at $\sqrt{s} = 6.6\,\text{GeV}$ at BNL,[93] at $\sqrt{s} = 19.4\,\text{GeV}$ at Fermilab,[91] and at $\sqrt{s} = 62.4\,\text{GeV}$ at RHIC[94] all show a striking similarity.

In the 1980s, a new 20 TeV on 20 TeV proton–proton collider was being planned in the US. Each 20 TeV SSC ring would have about 36,000 resonances. It was concluded that it should be possible to accelerate and maintain the polarization of 20 TeV protons in the SSC, only if the new Siberian-snake (named by Courant) concept of Derbenev and Kondratenko[95] really worked. This motivated the development at IUCF. Twenty six empty spaces for Siberian snakes were added in each SSC ring. The SSC was cancelled in 1993.

C.4.3 *Spin in storage rings*

During this period, considerable progress was made in the manipulation of spin in storage rings. In 1989, the Siberian Snake concept was demonstrated for the first time at IUCF[96] and plans proceeded to implement this technology in RHIC at BNL. In 1984, construction of the HERA electron–proton collider got underway with planned polarized e^-/e^+ beams using Sokolov–Ternov polarization and Richter-Schwitters spin-rotators.[97] In 1988, the HERMES experiment was proposed and, subsequently in 1994, significant lepton polarization was observed in HERA at 27 GeV via the Sokolov-Ternov mechanism for the first time.

The development of intense ($\approx 10^{17}$/s), highly polarized ($\approx 95\%$) atomic beams to feed storage cells made substantial progress in the 1980s.[98] In addition, a new generation of powerful lasers enabled the development of optically pumped sources of polarized hydrogen, deuterium and ^{3}He at ANL, Caltech, Harvard, and Princeton.[99] Work at Madison and Heidelberg on ABS fed internal gas targets was focused on a measurement of spin-filtering at the Heidelberg TSR.[100] At Novosibirsk, pioneering internal target experiments were carried out in collaboration with the ANL group.[101]

C.4.4 *Spin at medium energy accelerators*

In this period, a number of medium energy accelerators using polarized beams came online. Many were located at research universities and attracted new generations of young experimentalists into spin physics. These included Saclay in France, NIKHEF in the

Netherlands, Mainz and Bonn in Germany, TRIUMF in Canada, Novosibirsk in Russia, and MIT-Bates and IUCF in the US. With the decision to construct a new, CW, multi-GeV electron accelerator in the US, significant theoretical work began on electromagnetic nuclear physics. In particular, Donnelly and colleagues developed[102] the *SuperRosenbluth* technique, where spin was used as an experimental "knob" to maximize sensitivity to particularly important pieces of physics. This thinking had a major impact on the design of medium energy electron scattering experiments for the next several decades.

Scientific highlights from this period include: the first measurement in 1984 at MIT-Bates of t_{20} in elastic electron-deuteron scattering to allow separation of the three elastic form factors of the deuteron;[103] tests of the Standard Model via measurement of parity-violating quasielastic electron scattering from beryllium at Mainz[104] and elastic electron scattering from ^{12}C at Bates;[105] and the first measurements of spin-dependent electron scattering from polarized ^3He gas targets at MIT-Bates.[106] At IUCF in 1993, the first experiment to use both polarized beam (proton) and polarized internal gas target (^3He) was carried out[107] to study the spin structure of ^3He. The MIT-Bates and IUCF measurements showed that polarized ^3He could be used as an effective polarized neutron target for scattering experiments and set the stage for subsequent lepton scattering experiments at SLAC, DESY, Mainz, and Jefferson Lab.

Polarimetry, both for polarized beams, and for recoil particles, made enormous strides in this period. For electron beams, both Mott and Møller scattering were used as well as laser backscattering. Recoil neutron and proton polarimeters were developed to measure the nucleon elastic form-factors in the transfer of polarization from incident polarized electron beam on unpolarized targets. Subsequently, they have been used to great effect in the Jefferson Lab program.

C.4.5 *Hadronic parity violation*

Measurement of hadronic parity violation using spin observables became an active field of study in this period: see reviews in 1985,[108] in 1988[109] and in 2013.[110] The theoretical underpinning for this experimental program was the 1980 meson-exchange approach of Desplanques, Donoghue, and Holstein.[111] The analyzing power for the scattering of longitudinally polarized protons from an unpolarized proton target was measured at $13.6\,\text{MeV}$[112] and $15\,\text{MeV}$[113] by groups from Bonn and Los Alamos, and at $45\,\text{MeV}$[114] by a group from PSI. Further, a medium energy measurement at $220\,\text{MeV}$ was made at TRIUMF.[115] The TRIUMF experiment importantly used an optically pumped polarized proton source.[116] The measured asymmetry is typically of order 10^{-7}. Further experiments to find parity violation in $np \rightarrow d\gamma$ at Grenoble and LANL and in the circular polarization of $2.22\,\text{MeV}$ photons emitted in the capture of unpolarized thermal neutrons by protons at St. Petersburg were inconclusive. As so few of the experiments on NN and few body systems succeeded in isolating nonzero effects, experimenters turned to nuclei where parity violating observables might be enhanced. Several such experiments yielded non-zero results.

C.5 Spin in Use Worldwide: 1995–Present

Beginning in the mid-1990s, major experiments in spin physics at accelerators worldwide began to come online. They were built on the technical developments of previous decades and were motivated by important scientific questions. These included: the origin of nucleon spin; the use of the SuperRosenbluth technique to measure the elastic proton and neutron form-factors; and tests of the Standard Model.

For a modern review of the spin structure of the nucleon, see Aidala *et al.*[117] The HERMES experiment took data from 1995 to 2007, when operation of HERA ceased. It used the self-polarized $27\,\text{GeV}$ electron and positron beams of the HERA ring incident on polarized internal gas targets of hydrogen, deuterium,[118] and ${}^3\text{He}$.[119] For the first time, it employed sophisticated hadron detection and measured semi-inclusive DIS. HERMES provided first measurements of a number of important effects accessible in SIDIS or exclusive measurements. This has been subsequently pursued at Jefferson Lab and CERN/COMPASS and has underpinned the modern theoretical framework of nucleon structure based on multivariable Wigner distributions.

The COMPASS experiment continued the successful CERN polarized 160 GeV muon fixed target program into the new century. Both targets and detectors were substantially upgraded to include both longitudinal and transverse proton and deuteron as well as a RICH detector for particle identification. COMPASS provides unique data[120] at higher momentum transfer complementary to data from HERMES and Jefferson Lab.

With electron beam polarizations of over 80%, a new generation of experiments to measure inclusive DIS on polarized proton, deuteron, and ^{3}He targets in End Station A at SLAC yielded precision data on the spin-dependent structure functions.[121, 122] Targets based on atmosphere volumes of polarized ^{3}He were realized for the first time using spin exchange optical pumping[121] and used with 10 μA of electron beam. This technology has subsequently been further developed and employed to great effect at Jefferson Lab.

In the early part of this period, successful spin physics programs were carried out at the medium energy accelerators NIKHEF, IUCF, and MIT-Bates. At NIKHEF and Bates, stored electron beams of energy 600–850 MeV reached intensities of 250 mA and polarizations of about 65%. They were used with internal polarized gas targets to measure spin-dependent electron scattering from polarized hydrogen and deuterium.[123] A scientific highlight from this work was the determination of the neutron elastic electric form-factor at low momentum transfer with precision comparable to that of the proton.[124] At IUCF, the PINTEX program made precision measurements of spin-dependent proton–proton and proton–deuteron scattering between 100 and 500 MeV.[125] Also at IUCF, evidence for a three-nucleon force in spin-dependent elastic proton–deuteron scattering was obtained using a laser driven polarized deuterium target.[126] At both Bates and IUCF, highly efficient spin reversal of the stored beam was effected using an RF solenoid magnet.[127, 128] Regrettably, by 2005 the accelerators at all three laboratories had ceased operation. Fortunately, the COSY accelerator, with its polarized stored beams and polarized targets, remains an active laboratory for spin physics at medium energies.[129]

Although polarized beams were not in the original scope, by 1998 Jefferson Lab was operating at 4 GeV with intense, polarized electron beams. Spin measurements are central to Jefferson Lab's scientific program and the spin technical capabilities in terms of beams, targets, and polarimeters are world class. One of the major scientific results from Jefferson Lab was the discovery that the proton elastic form-factor ratio was dramatically different when measured with the recoil polarization technique compared to that obtained using the unpolarized cross-section.[130] The widely accepted explanation is that this is due to contributions beyond single photon exchange in the QED expansion.

Analysis of spin-dependent DIS data raised the possibility that the *strange* quarks carried unexpectedly large polarization. This motivated a worldwide program of parity violating electron scattering at MIT-Bates, Mainz, and Jefferson Lab to look for strange quark contributions to the proton's magnetic moment and charge.[131] While these experiments drove the technical development of polarized electron beams at these labs, by 2010 it was concluded that there were no large contributions from *strange* quarks.[132]

The RHIC accelerator turned on in 2000 with heavy ion beams and by 2006 the world's first polarized proton collider had been realized there. RHIC-spin employs a high intensity (3–5 mA), optically pumped >80% polarized proton source as well as Siberian Snakes and requires careful spin manipulation through multiple accelerators to achieve 60% polarized colliding proton beams at center-of-mass energies up to 500 GeV. It was made possible with substantial collaboration and investment from Japan. By 2011, the first measurement of parity-violating W-boson production was carried out.[133] In 2013, it was concluded that the contributions of gluons to the proton spin was comparable to that of the quarks.[134]

Polarization continues to be an essential experimental technique to test fundamental symmetries and to search for new physics beyond the Standard Model. At SLAC, experiment E158 carried out a precision test of the Standard Model by measuring the Weinberg angle in spin-dependent Møller scattering.[135] At BNL, E821 has reported a value for the muon's anomalous magnetic moment which challenges the Standard Model.[136] The availability of low-energy, polarized neutron beams for fundamental research has been realized. Polarized ^{3}He neutron spin filters are widely used[137] in this work. Searches for non-vanishing electric dipole moments (EDMs) of fundamental particles implies violation of both parity and time reversal symmetries. Several experiments to look for non-zero neutron EDM are under development.[138] Searches for non-zero EDMs of the proton and light nuclei in storage rings are also actively being considered.[139] Many exciting, open questions remain in spin physics.[140]

C.6 International Spin Physics Meetings

Research directed at understanding the spin of subatomic systems is an international enterprise where symposia and workshops are essential for progress. Beginning in 1960 in Basel, Switzerland, the International Symposia on Polarization Phenomena in Nuclear Reactions were initiated and subsequently took place every five years at Karlsruhe (1965), Madison (1970), Zürich (1975), Santa Fe (1980), Osaka (1985), and Paris (1990). Beginning in 1974 at Argonne National Laboratory, the International Symposia on High Energy Spin Physics were initiated and took place every two years at Argonne (1976, 78), Lausanne (1980), BNL (1982), Marseille (1984), Protvino (1986), Minneapolis (1988), Bonn (1990), and Nagoya (1992). In 1994, it was decided to merge the two symposia when the two meetings were held in parallel at Bloomington in that year. In 1996, the first International Spin Physics Symposium was held in Amsterdam and this was followed by biennial meetings at Protvino (1998), Osaka (2000), BNL (2002), Trieste (2004), Kyoto (2006), Charlottesville (2008), Jülich (2010), Dubna (2012), Beijing (2014), Urbana-Champaign (2016), and Ferrara (2018). The 2020 meeting was scheduled to take place in Matsue, Japan but was moved to 2021 because of the Covid-19 pandemic.

Fig. C.1: Group photo at the First International Spin Physics Symposium at Argonne National Laboratory in July 1974 to celebrate the world's first high energy polarized proton beam. From left to right: Herbert Anderson from U. Chicago and Fermi's chief assistant during the test of the first reactor in 1942; Paul Dirac; Robert Sachs, founder and director of Argonne National Laboratory; Louis Michel; Louis Dick from CERN and Alan Krisch from U. Michigan.

Since the mid-1960s there have been smaller, more focused workshops on spin tools and techniques. These have included an international meeting on: polarized targets and ion sources at Saclay in 1966, polarized targets at Berkeley in 1971, high intensity polarized proton sources at Ann Arbor in 1981 and at Vancouver in 1983, polarized target materials and techniques at Abingdon in 1981, at BNL in 1982, at Bad Honnef in 1984, polarized anti-protons at Bodega Bay in 1985, and polarized sources and targets at Montana, Switzerland in 1986, In 1988 and 1990, there were satellite workshops at the High Energy Spin Physics Symposia and in 1990 at KEK, Japan there was a workshop on polarized ion sources and gas jets. After 1990, these workshops were organized to take place in the uneven years between the international spin physics symposia. They took place in Heidelberg (1991), Madison (1993), Cologne (1995), Urbana (1997), Erlangen (1999), Nashville (2001), Novosibirsk (2003), Tokyo (2005), BNL (2007), Ferrara (2009), St. Petersburg (2011), Charlottesville (2013), Bochum (2015), KAIST (2017), and Knoxville (2019) Fig. C.1 shows a group of leading physicists at the First International Spin Physics Symposium in High Energy Physics at Argonne National Laboratory in July 1974.

These symposia and workshops are organized by the International Spin Physics Committee (ISPC). Alan Krisch (see Fig. 3.5) has played a leadership role in this committee since 1974. Starting in the 1980s, HERMES polarized targets, the HERA polarized beam and associated polarimeters have been a major focus at these international meetings. Further, the leadership of the ISPC since 2009 has been in the hands of HERMES physicists with Erhard Steffens (2010–2013), Richard Milner (2014–2017), Haiyan Gao (2018–2022) and Paolo Lenisa (2023–2026) serving as successive Chairs.

Bibliography

1. H. Schmidt-Boecking, Europhysicsnews EPN 50/3 (2019), p. 15.
2. P.A.M. Dirac, An historical perspective of spin, *Proceedings of the Summer Studies on High-energy Physics with Polarized Beams*, Argonne National Laboratory, July 1974.
3. G.E. Uhlenbeck, Fifty years of spin: Personal reminiscences, *Physics Today* (American Institute of Physics) **29**, 43 (1976).
4. S.A. Goudsmit, It might as well be spin, *Physics Today* (American Institute of Physics) **29**, 6 (1976).
5. G. Farmelo, *The Strangest Man* (a biography of Paul Dirac), Basic Books, 2011.
6. W. Haeberli, Proc. Pol. Phen. in Nucl. Physics, Karlsruhe 1965, P. Huber and H. Schopper (Eds.), Birkhäuser Verlag (1966), p. 64.
7. M.D. Barker *et al.*, *Proc. Pol. Phen. in Nucl. Physics*, Santa Fe 1980. *AIP Conf. Proceed.* **69**, 931 (1981).
8. E. Segré: Otto Stern 1888–1969, Biograph. Mem.
9. C. Grosshauser, Dipl. thesis, Erlangen 1994.
10. C. Diaconu, T. Haas, M. Medinnis, K. Rith, and A. Wagner, *Ann. Rev. Nucl. Part. Sci.* **60**, 101 (2010).
11. A. Airapetian *et al.* (The HERMES Collaboration), *Phys. Rev. D* **71**, 012003 (2005).
12. A. Airapetian *et al.* (The HERMES Collaboration), *Phys. Rev. Lett.* **87**, 182001 (2001).
13. A. Airapetian *et al.* (The HERMES Collaboration), *Nucl. Phys. B* **780**, 1 (2007).
14. K. Rith and A. Schäfer, *The Mystery of Nucleon Spin*, Scientific American, July 1999.
15. M.J. Tannenbaum, *Spin Physics at RHIC: A New Twist on the Heavy Ion Experiments*, BNL-63346, July 3, 1996.
16. T. Sloan, *History of the European Muon Collaboration (EMC)*, CERN Yellow Reports: Monograph, CERN-2019-005.

17. S.J. Brodsky, F. Fleuret, C. Hadjidakis, and J.P. Lansberg, Physics opportunities of a fixed-target experiment using LHC beams, *Phys. Reports* **522**, 239 (2013).
18. E. Steffens, Estimation of the performance of a HERMES-type gas target internal to the LHC, *Proc. Sci.*, PoS(PSTP2015)019.
19. AFTER@LHC, initiative for FT experiment at the LHC; http://after.in2p3.fr/after/index.php/Main_Page.
20. G.N. Lewis, *J. Am. Chem. Soc.* **38**, 762 (1916).
21. I. Langmuir, *J. Am. Chem. Soc.* **41**, 868 (1919).
22. W. Gerlach and O. Stern, *Zeitschrift für Physik* **9**, 353 (1922).
23. S.N. Bose, *Zeitschrift für Physik* **26**, 178 (1924).
24. W. Pauli, *Zeitschrift für Physik* **31**, 765 (1925).
25. G.E. Uhlenbeck and S. Goudsmit, *Naturwissenchaften* **13**, 953 (1925); *Nature* **117**, 264 (1926).
26. L.H. Thomas, *Nature* **117**, 514 (1926).
27. E. Fermi, *Rendiconti Lincei* **3**, 145 (1926).
28. P.A.M. Dirac, *Proc. Roy. Soc. A* **112**, 661 (1926).
29. R.H. Fowler, *Month. Noti. Roy. Astron. Soc.* **87**, 114 (1926).
30. A. Sommerfeld, *Naturwissenschaften* **15**, 824 (1927).
31. R.H. Fowler and L.W. Nordheim, *Proc. Roy. Soc. A* **119**, 173 (1928).
32. P.A.M. Dirac, *Proc. Roy. Soc. A* **117**, 610 (1928).
33. K. Wrede, *Zeitschrift für Physik* **41**, 569 (1927).
34. T.E. Phipps and J.B. Taylor, *Phys. Rev.* **29**, 309 (1927).
35. N. Mott, *Proc. Roy. Soc. A* **124**, 425 (1929).
36. C.G. Shull, C.T. Chase, and F.E. Myers, *Phys. Rev.* **63**, 29 (1943).
37. I. Estermann, R. Frisch, and O. Stern, *Nature* **132**, 169 (1933).
38. J.M.B. Kellogg, I.I. Rabi, N.F. Ramsey, Jr., and J.R. Zacharias, *Phys. Rev.* **56**, 728 (1939).
39. M.G. Mayer, *Phys. Rev.* **74**, 235 (1948).
40. T.D. Lee and C.N. Yang, *Phys. Rev.* **104**, 822, (1956).
41. C.S. Wu, E. Ambler, R.W. Hayward, D.D. Hoppes, and R.P. Hudson, *Phys. Rev.* **105**, 1413 (1957).
42. M. Goldhaber, L. Grodzins, and A.W. Sunyar, *Phys. Rev.* **109**, 1015 (1958).
43. V. Bargmann, L. Michel, and V.L. Telegdi, *Phys. Rev. Lett.* **2**, 435 (1959).
44. J.D. Bjorken, *Phys. Rev.* **148**, 1467 (1966).
45. J. Ellis and R. Jaffe, *Phys. Rev. D* **9**, 1444 (1974).
46. H. Paetz gen. Schieck, *Nuclear Physics with Polarized Particles*, 1st edn. (Lecture Notes in Physics Vol. 842), Springer, November 2011.

47. P. Huber and K.P. Meyer, eds. Proceedings of the International Symposium on Polarization Phenomena of Nucleons, *Helvetica Phys. Acta, Suppl. VI* (1961).

48. H.H. Barschall and W. Haeberli, eds. *Polarization Phenomena in Nuclear Reactions*, The University of Wisconsin Press, Madison (1971), p. XXV.

49. W. Haeberli, *Ann. Rev. Nucl. Sci.* **17**, 373 (1967).

50. Willy Haeberli, The toolbox of proton spin physics in historical perspective, PSTP2007, BNL, *AIP Conf. Proc.* **980**, p. 3 (2008).

51. J. Schwinger, Abstract of contributed paper B12 at the meeting of the *American Physical Society in Cambridge*, MA on April 25–27, 1946.

52. L. Wolfenstein, *Phys. Rev.* **75**, 1664 (1949).

53. M. Heusinkveld and George Freier, *Phys. Rev.* **85**, 80 (1952).

54. H.A. Bethe and F. de Hoffmann, *Mesons and Fields* **II**, 72 (Row, Peterson & Co., Evanston, Ill., 1955).

55. G. Clausnitzer, R. Fleischmann, and H. Schopper, *Z. Physik* **144**, 336 (1956).

56. A. Abragam and J.M. Winter, *Phys. Rev. Lett.* **1**, 374 (1958).

57. W. Gruebler, W. Haeberli, and P. Schwandt, *Phys. Rev. Lett.* **12**, 595 (1964).

58. C.Y. Prescott, *The Polarized Electron Story: From Atomic Beams to the International Linear Collider*, BNL, May 2006.

59. R.L. Long, Jr., W. Raith, and V.W. Hughes, *Phys. Rev. Lett.* **15**, 1 (1965).

60. A.A. Sokolov and I.M. Ternov, *Doklady Akademii Nauk SSSR* **153**, 1052 (1963).

61. J. Arianer *et al.*, *Nucl. Instr. Meth.* **A337**, 1 (1993).

62. C.D. Keith, *Dynamically Polarized Targets: A Historical Perspective*, Jefferson Lab presentation, 2013.

63. S. Bernstein *et al.*, *Phys. Rev.* **94**, 1243 (1954).

64. A. Overhauser, *Phys Rev.* **92**, 411 (1953).

65. T. Carver and C. Slichter, *Phys. Rev.* **92**, 212 (1953).

66. C.D. Jefferies, *Phys. Rev.* **106**, 164 (1957).

67. A. Abragam and W.G. Proctor, *Compt. Rend. Acad. Sci.* (Paris) **246**, 2253 (1958).

68. A. Abragam *et al.*, *Phys. Lett.* **2**, 310 (1962).

69. O. Chamberlain *et al.*, *Phys. Lett.* **7**, 293 (1963).

70. M. Borghini, in *Method in Subnuclear Physics*, Vol IV, Part 2, ed. M. Nikolii, Gordon and Breach, New York, 1970, p. 191.

71. M. Borghini, in *Proc. Int. Conf. on Polarized Targets*, ed. G. Shapiro, Berkeley, 1971, p. 1.
72. W. de Boer and T.O. Niinikoski, *Nucl. Inst. Meth.* **114**, 495 (1974).
73. T.O. Niinikoski and F. Udo, *Nucl. Inst. Meth.* **134**, 219 (1976).
74. T.O. Niinikoski, *Nucl. Instr. Meth.* **97**, 95 (1971).
75. T.O. Niinikoski and J.M. Rieubland, *Phys. Lett.* **72A**, 141 (1979).
76. W. Meyer, *Nucl. Inst. Meth. A* **526**, 12 (2004).
77. D.G. Crabb and W. Meyer, *Ann. Rev. Nucl. Part. Sci.* **47**, 67 (1997).
78. A. Kastler, *J. Phys.* **II**, 255 (1950).
79. F.D. Colegrove, L.D. Schearer, and G.K. Walters, *Phys. Rev.* **132**, 2561 (1963).
80. Ch. Leemann *et al.*, *Ann. Phys.* **66**, 810 (1971).
81. H.R. Weller and D.R. Lehman, *Ann. Rev. Nucl. Part. Sci.* **38**, 563 (1988).
82. V.W. Hughes and J. Kuti, *Ann. Rev. Nucl. Part. Sci.* **33**, 611 (1983).
83. J. Ashman *et al.*, *Phys. Lett. B* **206**, 364 (1988).
84. C.Y. Prescott *et al.*, *Phys. Lett. B* **84**, 524 (1979).
85. T. Maruyama *et al.*, *Phys. Rev. Lett.* **66**, 2376 (1991).
86. R.A. Assmann, Mike Lamont, and Steve Myers, *A Brief History of the LEP Collider.*
87. A.D. Krisch, *Eur. Phys. J. A* **31**, 417 (2007).
88. R.C. Fernow and A.D. Krisch, *Ann. Rev. Nucl. Part. Sci.* **31**, 107 (1981).
89. J.R. O'Fallon *et al.*, *Phys. Rev. Lett.* **39**, 733 (1977).
90. K. Heller *et al.*, *Phys. Rev. Lett.* **51**, 2025 (1983).
91. D.L. Adams *et al.*, *Phys. Lett. B* **264**, 462 (1991).
92. R.D. Klem *et al.*, *Phys. Rev. Lett.* **36**, 929 (1976).
93. C.E. Allgower *et al.*, *Phys. Rev. D* **65**, 092008 (2002).
94. I. Arsene *et al.*, *Phys. Rev. Lett.* **101**, 042001 (2008).
95. Ya.S. Derbenev and A.M. Kondratenko, *Sov. Phys. Dokl.* **20**, 562 (1978).
96. A.D. Krisch *et al.*, *Phys. Rev. Lett.* **63**, 1137 (1989).
97. R. Schwitters and B. Richter, *PEP Note* **87** (1974).
98. E. Steffens and W. Haeberli, *Rep. Prog. Phys.* **66**, 1887 (2003).
99. T.E. Chupp, R.J. Holt, and R.G. Milner, *Ann. Rev. Nucl. Part. Sci.* **44**, 373 (1994).
100. F. Stock *et al.*, *Nucl. Instr. Meth.* **A343**, 334 (1994).
101. B.B. Wojtsekhowski *et al.*, *JETP Lett.* **43**, 733 (1986); R. Gilman *et al.*, *Phys. Rev. Lett.* **65**, 1733 (1990).
102. T.W. Donnelly and A.S. Raskin, *Ann. Phys.* (NY) **169**, 247 (1986).
103. I. The *et al.*, *Phys. Rev. Lett.* **67**, 173 (1991).
104. W. Heil *et al.*, *Nucl. Phys. B* **327**, 1 (1989).

105. P.A. Souder *et al.*, *Phys. Rev. Lett.* **65**, 694 (1990).
106. C.E. Woodward *et al.*, *Phys. Rev. Lett.* **65**, 698 (1990); A.K. Thompson *et al.*, *Phys. Rev. Lett.* **68**, 2901 (1992); H. Gao *et al.*, *Phys. Rev. C* **50**, R546 (1994); J.-O. Hansen *et al.*, *Phys. Rev. Lett.* **74**, 654 (1995).
107. K. Lee *et al.*, *Phys. Rev. Lett.* **70**, 738 (1993).
108. E.G. Adelberger and W.C. Haxton, *Ann. Rev. Nucl. Part Sci.* **35**, 501 (1985).
109. See contributions in *Proc. Workshop on Parity Violation in Hadronic Systems, Can. J. Phys.* **66**(6), 485–567 (1988).
110. W.C. Haxton and B.R. Holstein, arXiv:1303.4132.
111. B. Desplanques, J.F. Donoghue, and B.R. Holstein, *Ann. Phys.* (NY), **124**, 449 (1980).
112. P.D. Eversheim *et al.*, *Phys. Lett.* **B256**, 11 (1991).
113. D.E. Nagle *et al.*, *AIP Conf. Proc.* **51**, 224 (1978).
114. R. Balzer *et al.*, *Phys. Rev. Lett.* **44**, 699 (1980).
115. A.R. Berdoz *et al.*, *Phys. Rev. Lett.* **87**, 272301 (2001).
116. C.D.P. Levy *et al.*, *Proc. ICIS'97, Whistler, Rev. Sci. Instr.* **67**, 1291 (1996).
117. C.A. Aidala, S.D. Bass, D. Hasch, and G.K. Mallot, *Rev. Mod. Phys.* **85**, 655 (2013).
118. A. Airapetian *et al.*, *Nucl. Instr. Meth. A* **540**, 68 (2005).
119. D. DeSchepper *et al.*, *Nucl. Instr. Meth. A* **419**, 16 (1998).
120. A. Martin, Contribution to SPIN2012, Dubna, Russia, September 20–24, 2012.
121. P.L. Anthony *et al.*, *Phys. Rev. D* **54**, 6620, (1996).
122. P.L. Anthony *et al.*, *Phys. Lett. B* **553**, 18, 2003.
123. D.K. Hasell *et al.*, *Ann. Rev. Nucl. Part. Sci.* **61**, 409 (2011).
124. E. Geis *et al.*, *Phys. Rev. Lett.* **101**, 042501 (2008).
125. B. v. Przewoski *et al.*, *Phys. Rev. C* **74**, 064003 (2006).
126. R.V. Cadman *et al.*, *Phys. Rev. Lett.* **86**, 967 (2001).
127. B.B. Blinov, *Phys. Rev. Lett.* **81**, 2906 (1998).
128. V.S. Morozov *et al.*, *Phys. Rev. ST A* **4**, 104002 (2001).
129. F. Rathmann, *J. Phys. Conf. Ser.* **295**, 012006 (2011).
130. A.J.R. Puckett *et al.*, *Phys. Rev. C* **85**, 045203 (2012).
131. D.H. Beck and R.D. McKeown, *Ann. Rev. Nucl. Part. Sci.* **51**, 189 (2001).
132. D.S. Armstrong and R.D. McKeown, *Ann. Rev. Nucl. Part. Sci.* **62**, 337 (2012).
133. M.M. Aggarwal *et al.*, *Phys. Rev. Lett.* **106**, 062002 (2011).

134. B. Surrow, Contribution to QCD Evolution Workshop (QCD2013), Jefferson Laboratory, Newport News, VA, May 6–10, 2013.
135. P.L. Anthony *et al.*, *Phys. Rev. Lett.* **95**, 081601 (2005).
136. G.W. Bennett *et al.*, *Phys. Rev. D* **73**, 072003 (2006).
137. M.G. Huber *et al.*, *Phys. Rev. Lett.* **102**, 200401 (2009).
138. T.M. Ito, *J. Phys. Conf. Ser.* **69**, 012037 (2007).
139. P. Lenisa, Contribution to the Proceedings of the XVth International Workshop on Polarized Sources, Targets and Polarimetry, U. of Virginia, Charlottesville, USA, September 2013.
140. R.L. Jaffe, *J. Mod. Phys. A* **18**, 1141 (2003).

Lightning Source UK Ltd.
Milton Keynes UK
UKHW051418260622
404964UK00001B/14